532582

D1685738

This book is to be returned on or before
the last date stamped below.

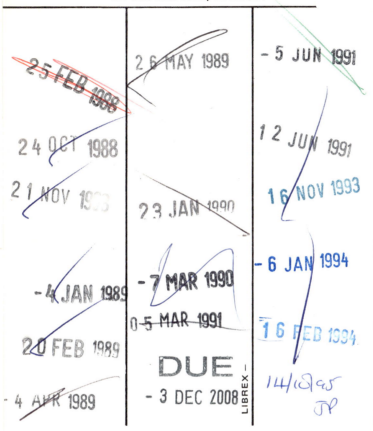

25 FEB 1988

2 6 MAY 1989

- 5 JUN 1991

24 OCT 1988

1 2 JUN 1991

2 1 NOV 19

2 3 JAN 1990

1 6 NOV 1993

- 4 JAN 1989

- 7 MAR 1990

- 6 JAN 1994

2 0 FEB 1989

0 5 MAR 1991

1 6 FEB 1994

- 4 APR 1989

DUE
- 3 DEC 2008

LIBREX —

14/10/95
JP

ELECTRONIC NOISE

McGRAW-HILL
INTERNATIONAL
BOOK COMPANY

New York
St. Louis
San Francisco
Auckland
Bogotá
Guatemala
Hamburg
Johannesburg
Lisbon
London
Madrid
Mexico
Montreal
New Delhi
Panama
Paris
San Juan
São Paulo
Singapore
Sydney
Tokyo
Toronto

ANDRÁS AMBRÓZY

Department of Electronics Technology
Technical University, Budapest

Electronic Noise

The original:

ELEKTRONIKUS ZAJOK

was published by Műszaki Könyvkiadó, Budapest

Translated by

T. SÁRKÁNY

British Library Cataloguing in Publication Data

Ambrózy, András
  Electronic noise
  1. Electronic noise
  I. Title
  621.38′0436   TK7867.5   80-40313
  ISBN 0-07-001124-9

Joint edition published by McGraw-Hill International Book Company, New York, U.S.A. and Akadémiai Kiadó, The Publishing House of the Hungarian Academy of Sciences, Budapest, Hungary

ELECTRONIC NOISE

1 2 3 4 8432

Printed and bound in Hungary

CONTENTS

Theoretically, the unlimited amplification of electrical signals, i.e. the detection of arbitrarily small signals, has been implemented by the vacuum tube which may be considered to be the first active element of electronics. However, it soon became evident that these small signals are absorbed by disturbances of various origins. In the early years of radio reception, atmospheric and power line interferences were observed; cosmic noise sources have been discovered subsequently. However, in spite of perfect screening against external disturbances, random small voltage and current fluctuations in circuits have been experienced. Investigations have shown that these fluctuations have two origins.

1. Current is carried by elementary charges having uneven energy distribution. This has the effect that the number of charges per unit time passing through the vacuum space of an electron tube or the p–n junction of a solid state device is not constant.
2. A random brownian motion by elementary charges results in collision between these charges and the atoms of the solid body forming the circuit. Each collision has the effect of changing the direction and magnitude of an elementary current filament.

Both phenomena are outcomes of fundamental laws of nature so the signal fluctuations caused by these phenomena even theoretically cannot be eliminated. The term "electronic noise", in short "noise", will be applied to describe these phenomena. This book deals exclusively with noise; and external disturbances will not be covered.

During the evolution of radio communication, with increasing annoyance, noise became a limiting factor. A thorough investigation of noise was therefore launched leading to unexpected, useful results. It turned out that a knowledge of noise properties may contribute to our understanding of the physical background of active device operation. Very soon, in addition to theoretical investigations, the evolution of noise-measuring methods of increasing sensitivity and accuracy gained in importance.

Theoretical investigations have shown that noise may be considered as a series of random stochastic events. Moreover, measurements have proved that in most cases the useful signals to be transmitted also have stochastic properties. It is thus appropriate to investigate electronic devices, circuits, and systems by applying stochastic driving signals. The response of a system driven by a stochastic signal will include the responses to sinusoidal and step signals, as treated by classical electrical theorems.

The methods of mathematical statistics and probability theory, taking shape during this period, have turned out to offer useful means of analyzing random signals. However, these methods in themselves seemed to be insufficient, and several additional considerations were needed to analyze and understand basically the physical origins of noises appearing in electronic devices and circuits. The practicing engineer may have problems in trying to form a unified picture based on theoretical physical, mathematical, and his own practical considerations.

This book's main task is to help the engineer in his composite model-forming process. The extensive literature available may provide a suitable background, but a definite starting point and guidance is needed for the correct evaluation of highly specialized references. The definite statement of basic relations is only possible if preceded by the formation of a unified system comprising basic background knowledge and the physical contents of the mathematical formulas.

The main relations refer to noise power. With due caution, these may be extended to voltages and currents, and in the case of linear circuits, linear superposition may be applied to these voltages and currents. Noise equivalent circuits may be introduced, leading to well-proven circuit design methods. These are then followed by noise parameters and noise equivalent circuits of active devices.

Noise measurements usually require a detector having some kind of nonlinear characteristic. This requires, to some extent, the analysis of responses resulting from noise driven, nonlinear networks. The possible methods of noise measurement will also be investigated, and the theoretical limitations of noise measurements will be treated.

This relatively short book cannot go into details of all the existing problems. For more detailed investigations, the reader is referred to the references at the end of each Chapter.

András Ambrózy

ONE

INTRODUCTION

As an introduction, a few important basic terms will be defined by the analysis of a simple example, the shot-effect noise. However, this analysis cannot yet be complete as information treated in later Chapters is not yet known. We shall come back to the unanswered questions in the appropriate Chapters.

1-1 BASIC TERMS

Figure 1-1 shows the energy diagram of a vacuum-tube cathode and of a p–n junction. In order that an electron can pass from the cathode into the vacuum or from the n layer into the p layer, it must possess sufficient energy to pass through the energy gap. The energies of electrons are randomly distributed, resulting in a fluctuation of charges passing over per time unit, and thus in a fluctuation of current. In a loudspeaker, the sound of the amplified current fluctuation is similar to the dropping of shots, explaining the term shot noise.

In order to calculate this fluctuation, let us investigate an interval of time T, and let us assume that during this interval the number of electrons passing through the energy gap will be N. Let us further assume that at the beginning of this time interval, the number of electrons within the cathode (or within the n layer) will also be N, and for the sake of our mental experiment, let us distinguish between these electrons in some way. After the time interval T, these electrons will be used up.

Let us now divide the time interval investigated into intervals Δt of equal lengths. As shown in Fig. 1-2, the number of electrons passing through the energy gap during intervals $\Delta t_1, \Delta t_2, ..., \Delta t_i$ will be $n_1, n_2, ..., n_i$. We note that $n_1, n_2, ..., n_i$ is not a monotonous series, members of different subscripts may be equal to each other. The following question should be answered: what is the probability of the electron bearing number 17, for example, passing through the energy gap in that time interval which is lined up in Fig. 1-2?

This event would certainly take place during time interval T, so the probability of it occurring during Δt is $\Delta t/T$. The situation is different if the probability of a definite electron combination passing (e.g., electrons 1, 17, and 23) has to be calculated (see Fig. 1-1). During this time interval, only the designated combination is allowed to pass over, excluding all those electrons which do not belong to the combination. The probability of not passing will be $1-\Delta t/T$ for each single electron. The movements of electrons are independent of each other. The joint probability of occurrence of independent events is given by the

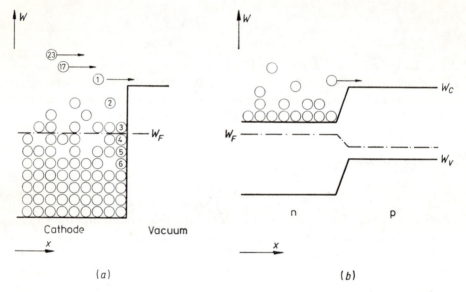

Figure 1-1 *(a)* Energy gap in a vacuum tube. *(b)* Energy gap in a p–n junction. (W, electron energy; W_C, bottom of the conduction band; W_F, Fermi level, W_V, top of the valence band).

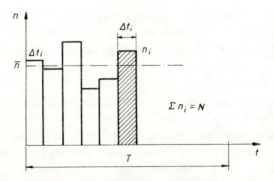

Figure 1-2 Model of shot-noise origin.

product of the single probabilities of occurrences, thus the passing probability of a particular combination is

$$\left(\frac{\Delta t}{T}\right)^{n_i}\left(1-\frac{\Delta t}{T}\right)^{N-n_i}$$

However, single electrons may not actually be distinguished. What may be stated is only: what is the probability of the joint passing of n_i electrons during an interval Δt? From a group of N electrons there are $\binom{N}{n_i}$ ways to select n_i electrons. Thus the probability of the above joint passing will be stated by the well-known formula giving the probabilities of events occurring according to the binomial distribution. Thus

$$f(n_i) = \binom{N}{n_i}\left(\frac{\Delta t}{T}\right)^{n_i}\left(1-\frac{\Delta t}{T}\right)^{N-n_i} \tag{1-1}$$

This distribution type is characteristic of cases in which alternatives are governed by chance. (In our case the alternatives are given by "pass occurs"—"pass does not occur".) However, it is rather inconvenient to apply the binomial distribution, especially for very large numbers of N. A good approximation is achieved by allowing both N and T to increase beyond any limit, i.e. $T\rightarrow\infty$, $N\rightarrow\infty$, and thus $\Delta t/T\rightarrow 0$, but

$$N\frac{\Delta t}{T} = \lambda \tag{1-2}$$

where λ is a finite and constant number. It is immediately evident from Fig. 1-2 that Eq. (1-2) gives the average number of electrons passing during time interval Δt. Taking into account Eq. (1-2) in rewriting Eq. (1-1), we have

$$f(n) = \binom{N}{n}\left(\frac{\lambda}{N}\right)^{n}\left(1-\frac{\lambda}{N}\right)^{N-n}$$

$$= \frac{\lambda^{n}}{n!}\frac{N(N-1)...(N-n+1)}{N^{n}}\left(1-\frac{\lambda}{N}\right)^{-n}\left(1-\frac{\lambda}{N}\right)^{N} \tag{1-3}$$

The second factor in the right-hand four factor product approaches 1 if $N\rightarrow\infty$. Thus

$$\lim_{N\rightarrow\infty}\left[1\left(1-\frac{1}{N}\right)\left(1-\frac{2}{N}\right)...\left(1-\frac{n-1}{N}\right)\right] = 1$$

The third factor is

$$\lim_{N\rightarrow\infty}\left[\left(1-\frac{\lambda}{N}\right)^{-n}\right] = 1$$

and the fourth factor is

$$\lim_{N\rightarrow\infty}\left[\left(1-\frac{\lambda}{N}\right)^{N}\right] = e^{-\lambda}$$

From these expressions, we have

$$\lim_{\substack{N\to\infty \\ \Delta t/T\to 0}} f(n) = \frac{\lambda^n}{n!} e^{-\lambda} \tag{1-4}$$

which is the Poisson distribution as a limiting case of the binomial distribution. The probability of the passing of n electrons during time interval dt is $f(n)\,dt$. In spite of the limiting process $N\to\infty$, the distribution is not continuous and is interpreted for only positive integers of n. A fractional number of electrons cannot be interpreted physically.

This approximation of the binomial distribution can only be applied for cases of $\Delta t/T \ll 1$. This condition is not met for every fluctuation phenomenon!

If the fluctuation of n may be described by the Poisson distribution, it may be shown that the mean value of n is

$$M(n) \equiv \bar{n} = \lambda \tag{1-5}$$

where the mean value is denoted by $M(\ )$ (see Chapter 2). On the other hand, the variance of n, i.e. the mean square value of the fluctuation, is

$$D^2(n) \equiv \overline{(n-\bar{n})^2} = \lambda \tag{1-6}$$

Let us now calculate the current carried by the electrons. The instantaneous value of the current* in a time interval Δt is

$$i = \frac{nq}{\Delta t} \tag{1-7}$$

where $|q| = 1.6 \times 10^{-19}$ As, the elementary charge. The mean value of the current is

$$I = \bar{i} = \frac{\bar{n}q}{\Delta t} \tag{1-8}$$

The mean square value of the current fluctuation, which is directly proportional to the noise power, is given by

$$\overline{i^2} = \overline{(i-I)^2} = D^2(i) = \left(\frac{q}{\Delta t}\right)^2 D^2(n) = \left(\frac{q}{\Delta t}\right)^2 \bar{n} \tag{1-9}$$

Expressing $\bar{n}$ from Eq. (1-8) and substituting into Eq. (1-9) we have

$$\overline{i^2} = \frac{q}{\Delta t} I \tag{1-10}$$

---

* If the DC component is also taken into account in the instantaneous value, the symbol i is used. The DC mean value is denoted by upper case letters, the stochastic signal of zero mean value is denoted by lower case letters.

Thus the mean square value of the current fluctuation is directly proportional to the DC mean value and to the charge carrying the current, and is inversely proportional to the density of time interval divisions. The latter may be demonstrated in Fig. 1-3. The right-hand side shows the same fluctuation phenomenon as the left-hand side, but the intervals Δt_1 have been united in groups of five, thus smoothing out the short-lasting fluctuations.

Δt may not be chosen arbitrarily because this would not allow a unique definition of $\overline{i^2}$. On the other hand, should we consider the stochastic time function shown in Figs. 1-2 and 1-3, respectively, as a series of pulses and should we pass these pulses through a linear network, then the duration of the shortest pulse that could be transmitted through this network would be of the order of the reciprocal bandwidth. Thus the mean square current fluctuation is a function of the bandwidth:

$$\boxed{\overline{i^2} = 2qI\Delta f}$$
(1-11)

The conversion from Eq. (1-10) to Eq. (1-11) is not explained here, but we shall come back to this at the end of Chapter 3.

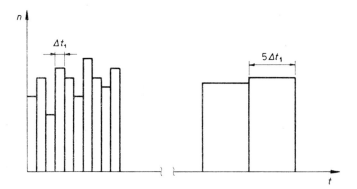

Figure 1-3 Changing of time division.

The frequency is not contained in Eq. (1-11), thus the frequency band of width Δf may be selected anywhere. This infers that $\overline{i^2} \to \infty$ would result in the case of $\Delta f \to \infty$, which is evidently impossible. In Chapter 3, it will be shown that the finite time needed for the passage (e.g., the electron transit time in a vacuum tube) results in a frequency-dependent factor in Eq. (1-11) of less than one.

The shot noise is only an example of fluctuation phenomena in electronic devices. In order to achieve a unified treatment a few general terms are now introduced.

1-2 AMPLITUDE DENSITY

The instantaneous value of the current flowing through an electronic device may be given in the form

$$i(t) = I + i(t) \tag{1-12}$$

which is similar to that used for periodic signals superimposed on direct current. Unfortunately, this similarity is of no use here as the exact function $i(t)$ is not known. On the other hand, the probability of n electrons participating in the current flow may be given using Eq. (1-4)

$$f(n) = \frac{\lambda^n}{n!} e^{-\lambda}$$

In practical cases, n is a very large number. When utilizing current-measuring instruments, we can not distinguish single electrons, only finite current ranges. The probability of $i(t)$ falling between I_0 and $I_0 + \Delta I$ if $\Delta I \to 0$ (see Fig. 1-4a) is by

$$\mathscr{P}[I_0 \leq i(t) < I_0 + \Delta I] = f(I_0)\Delta I \tag{1-13}$$

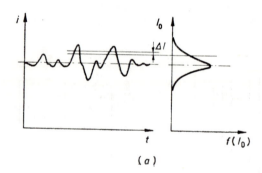

(a)

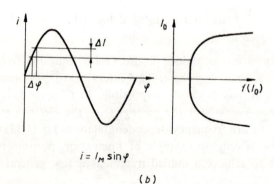

$$i = I_M \sin \varphi$$

(b)

Figure 1-4 Time function and amplitude density function. *(a)* Stochastic signal. *(b)* Sine-wave signal.

The function $f(I_0)$ is called the amplitude density function, its dimension being A^{-1}. In a similar way, the amplitude density function $f(U_0)$ may also be derived, its dimension being V^{-1}.

In case of shot-effect noise, $f(I_0)$ is also a Poisson function as i is proportional to n. Evidently other kinds of distributions would be valid for other types of noise. The distribution types frequently occurring in the theory of electronic noise will be surveyed in Chapter 2.

Signals with known time behavior—i.e. deterministic signals—also have amplitude density functions. For instance, the function of the sinusoidal signal is shown in Fig. 1-4b. It is easily shown that

$$f(I_0)_{\sin} \Delta I = \frac{2\Delta\varphi}{2\pi} \tag{1-14}$$

and

$$\lim_{\Delta i \to 0} f(I_0)_{\sin} = \frac{1}{\pi} \frac{d\varphi}{di} = \frac{1}{\pi} \frac{d}{di}\left(\arcsin\frac{i}{I_{max}}\right) = \frac{1}{\pi I_{max}} \frac{1}{\sqrt{1-(I_0/I_{max})^2}} \tag{1-15}$$

A knowledge of amplitude density functions pertaining to deterministic signals is especially important for the investigation of nonlinear networks (e.g., demodulators).

1-3 POWER DENSITY

If the current having a mean square value calculated from Eq. (1-11) flows through a resistance we have a noise power of

$$P = \overline{i^2}R = 2qIR\Delta f \tag{1-16}$$

The power per unit bandwidth (power/Hz) is called power density. In case of shot-effect noise, this is given by

$$p = 2qIR\left(\frac{W}{Hz}\right) \tag{1-17}$$

If the bandwidth of the measuring instrument is limited to the low-frequency range, then the power density of the shot-effect noise is independent of frequency (see Fig. 1-5a). This means that the power spectrum is uniform, which is true in many practical cases. This kind of noise is called white noise, taking the analogy from the mixed nature of white light. However, in the general case (see Fig. 1-5b), we have

$$p = p(f) \quad \text{or} \quad p = p(\omega) \tag{1-18}$$

and the total noise is given by the integral

$$P = \int_0^\infty p(f)\,df \tag{1-19}$$

In practical calculations, frequently $\overline{i^2}$ or $\overline{u^2}$ are calculated. In most cases, it would be inconvenient to convert these quantities into powers. This is why the spectral density of the squared fluctuations is written in the form

$$S = S(f) \quad \text{or} \quad S = S(\omega) \tag{1-20}$$

(dimensions: A^2/Hz, V^2/Hz, $A^2/rad\ s^{-1}$, $V^2/rad\ s^{-1}$)

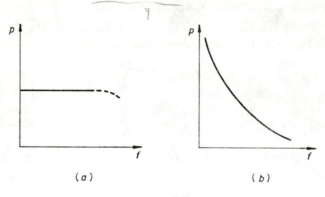

(a)   (b)

Figure 1-5 *(a)* Frequency-independent power spectrum (white noise). *(b)* Frequency dependent power spectrum (1/*f* noise).

Due to further considerations of convenience, the root mean square noise current and noise voltage

$$i_{rms} = \sqrt{\overline{i^2}} \qquad u_{rms} = \sqrt{\overline{u^2}} \tag{1-21}$$

are used. It should be stressed, however, that for stochastic signals, only spectra of the type given in Eqs. (1-18) and (1-20) may be interpreted. From these spectra, no conclusions regarding the frequency spectrum of the stochastically changing current or voltage may be drawn. On the other hand, the power spectrum may be easily determined from the so-called autocorrelation function dealt with in Chapter 3, which reveals an internal property of the stochastic signal.

PROBLEMS

1-1 Let the number of electrons passing during the time interval Δt be $\lambda = 10$. Let us compare the passing probabilities of 12 and 10 electrons!

Answer: From Eq. (1-4) we have

$$\frac{f(12)}{f(10)} = \frac{\lambda^{n_2} n_1!}{\lambda^{n_1} n_2!} = \frac{10^{12}}{10^{10}} \times \frac{1}{11 \times 12} = 0.76$$

Let us repeat this by taking $n=\lambda=100$, $n_2=120$, and $n_1=100$! For such large numbers, the values of factorials may be computed from the Stirling formula.

$$n! \cong n^n e^{-n} \sqrt{2\pi n} \qquad \text{if} \quad n \to \infty$$

$$f(100) = \left(\frac{\lambda}{\bar{n}}\right)^{n_1} \frac{e^{n_1}}{\sqrt{2\pi n_1}} e^{-\lambda} = \frac{e^{n_1}}{\sqrt{2\pi n_1}} e^{-\lambda}$$

$$(\lambda \equiv \bar{n})$$

$$f(120) = \left(\frac{\lambda}{n_2}\right)^{n_2} \frac{e^{n_2}}{\sqrt{2\pi n_2}} e^{-\lambda}$$

$$\frac{f(120)}{f(100)} = \left(\frac{1}{1.2}\right)^{120} e^{20} \sqrt{\frac{1}{1.2}} = 0.134$$

It can be seen that by increasing n, which according to Eq. (1-8) may be attained by increasing I or $\varDelta t$, the probability of a given relative difference, i.e. the relative fluctuation, gets smaller.

1-2 The power density of the so-called $1/f$ noise may be written in the form $p(f)=p_0 f_0/f$. Let $p_0=10^{-6}$ W/Hz and $f_0=1$ kHz. What is the noise power between 1 Hz and 10 kHz?

Answer:

$$P = p_0 f_0 \int_{f_1}^{f_2} \frac{df}{f} = p_0 f_0 \ln \frac{f_2}{f_1} = 9.21 \text{ mW}$$

Note that for $f_1 \to 0$, $P \to \infty$. A peculiarity of the $1/f$ noise is the fact that it follows the above frequency dependence even at frequencies several orders less than 1 Hz. At the same time, its power is finite, which may be explained by energetic reasoning. No satisfactory explanation for this phenomenon has yet been found.

BIBLIOGRAPHY

Books dealing with electronic noise and stochastic processes

[1] Bell, D. A.: *Electrical Noise*, D. Van Nostrand, London, 1960.
[2] Bendat, J. S.: *Principles and Applications of Random Noise Theory*, John Wiley, New York, 1958.
[3] — and A. G. Piersol: *Measurement and Analysis of Random Data*, John Wiley, New York, 1966, second revised edition, 1971.
[4] — and —: *Random Data. Analysis and Measurement Procedures*, Wiley Interscience, New York, 1971.
[5] Bennett, W. R.: *Electrical Noise*, McGraw-Hill, New York, 1960.
[6] Bittel, H. and L. Storm: *Rauschen*, Springer, Heidelberg, 1971.

[7] Bull, C. S.: *Fluctuations of Stationary and Nonstationary Electrical Current*, Butterworth, London, 1966.

[8] Burgess, R. E.: *Fluctuation Phenomena in Solids*, Academic Press, New York, 1965.

[9] Davenport, W. B. and W. L. Root: *An Introduction to the Theory of Random Signals and Noise*, McGraw-Hill, New York, 1958.

[10] Freeman, J. J.: *Principles of Noise*, John Wiley, New York, 1958.

[11] Harris, E. W. and T. J. Ledwidge: *Introduction to Noise Analysis*, Pion, London, 1974.

[12] King, R.: *Electrical Noise*, Chapman and Hall, London, 1966.

[13] Korn, G. A.: *Random Process Simulation and Measurements*, McGraw-Hill, New York, 1966.

[14] Lange, F. H.: *Korrelationselektronik*, Verlag Technik, Berlin, 1962.

[15] McDonald, D. K. C.: *Noise and Fluctuations*, John Wiley, New York, 1962.

[16] Newland, D. E.: *Random Vibrations and Spectral Analysis*, Longman, London, 1975.

[17] Papoulis, A.: *Probability, Random Variables, and Stochastic Processes*, McGraw-Hill, New York, 1965.

[18] Pfeiffer, H.: *Elektronisches Rauschen*, Teubner, Leipzig, vol. 1, 1959; vol. 2, 1968.

[19] Rice, S. O.: "Mathematical Analysis of Random Noise", in N. Wax (ed.), *Selected Papers on Noise*, Dover, New York, 1954.

[20] Robinson, F. N. H.: *Noise in Electrical Circuits*, Oxford University Press, Oxford, 1962.

[21] —: *Noise and Fluctuations in Electronic Devices and Circuits*, Clarendon Press, Oxford, 1974.

[22] Smullin, L. D. and H. A. Haus: *Noise in Electron Devices*, John Wiley, New York, 1959.

[23] van der Ziel, A.: *Noise*, Prentice Hall, Englewood Cliffs, 1954.

[24] —: *Fluctuation Phenomena in Semiconductors*, Butterworth, London, 1959.

[25] —: *Noise: Sources, Characterization, Measurement*, Prentice-Hall, Englewood Cliffs, 1970.

[26] —: *Noise in Measurements*, Wiley Interscience, New York, 1976.

Tutorial papers, conference proceedings:

[27] Bennet, W. R.: "Methods for Solving Noise Problems", *Proc. IRE*, vol. 44, no. 5, pp. 609—637, 1956.

[28] Gupta, M. S.: "Applications of Electrical Noise", *Proc. IEEE*, vol. 63, no. 7, pp. 996—1010, 1975.

[29] Johnson, J. B.: "Electronic Noise: The First Two Decades", *IEEE Spectrum*, vol. 8, no. 2, pp. 42—46, 1971.

[30] van der Ziel, A.: "Noise in Solid-State Devices and Lasers", *Proc. IEEE*, vol. 58, no. 8, pp. 1178—1206, 1970.

[31] *Conference on Physical Aspects of Noise in Electronic Devices, 1968, University of Nottingham*, Peter Peregrinus, Stevenage, Hertfordshire, 1968.

[32] "Le bruit de fond des composants actifs semiconducteurs": *Colloques Internationaux du Centre National de la Recherche Scientifique* (CNRS), no. 204, Toulouse, 1971.

[33] *Proceedings of the Fourth International Conference on Physical Aspects of Noise in Solid-State Devices*, Noordwijkerhout, 1975.

[34] Institute of Technology, *Symposium on 1/f Fluctuations*, Tokyo, 1977.

[35] Wolf, D. (ed.): *Proceedings of the Fifth International Conference on Noise in Physical Systems, Bad Nauheim, 1978*, Springer, Berlin, 1978.

Books dealing with the interfering effects
of noises on transmission
and with signal-to-noise ratio improvement:

[36] Blachman, N. M.: *Noise and Its Effect on Communication*, McGraw-Hill, New York, 1966.

[37] Goldman, S.: *Information Theory*, Prentice-Hall, Englewood Cliffs, 1953.

[38] Harman, W.: *Principles of the Statistical Theory of Communication*, McGraw-Hill, New York, 1963.

[39] Kotel'nikov, V. A.: *The Theory of Optimum Noise Immunity*, McGraw-Hill, New York, 1959.

[40] Lange, F. H.: *Signale und Systeme*, Verlag Technik, Berlin, 1965.

[41] Lee, Y. W.: *Statistical Theory of Communication*, John Wiley, New York, 1960.

[42] Motchenbacher, C. D. and F. C. Fitchen: *Low Noise Electronic Design*, John Wiley, New York, 1973.

[43] Schwartz, M.: *Information, Transmission, Modulation and Noise*, 2nd ed., McGraw-Hill, New York, 1970.

[44] — and Schaw, L.: *Signal Processing: Discrete Spectral Analysis, Detection and Estimation*, McGraw-Hill, New York, 1975.

TWO

DISTRIBUTION AND DENSITY FUNCTIONS OF INSTANTANEOUS VALUES

As mentioned in Chapter 1, the exact time function of a stochastic signal is unknown. However, the occurrence probabilities of given instantaneous values may be computed utilizing the physical background knowledge of the random process. The probability terms applied in our investigations will be treated without striving for complete coverage, and occasionally limited to our special needs. For thorough investigations, publications dealing with probability theory should be consulted.[4,5,11,12,14,15] However, care will be taken in making our relations suitable for describing the random variables of physical processes.

2-1 BASIC TERMS OF PROBABILITY THEORY

In a general sense, a random trial is characterized by the fact that its outcome is not determined uniquely by the conditions taken into account. The possible outcomes of a trial are called the elementary events. In a mathematical sense, a possible event is a subset of the set Ω comprising the elementary events forming the event space.

In probability theory, the Kolmogorov set of axioms is generally accepted. According to this set of axioms, the $P(A)$ function defined on the subsets of the event space Ω is called probability if the following axioms are met for this function:

$$0 \leq \mathscr{P}(A) \leq 1 \tag{2-1}$$

$$\mathscr{P}(\Omega) = 1 \tag{2-2}$$

$$\mathscr{P}\left(\sum_k A_k\right) = \sum_k P(A_k) \tag{2-3}$$

if $A_1, A_2, ..., A_k$ is a finite or infinite series of mutually exclusive events.

The random variable is a real function defined over the set Ω comprising the elementary events. The real number characterizing the elementary event ω will generally be denoted by ξ, and in some cases, the event ω will also be indicated, e.g., $\xi(\omega)$. In most cases, the numerical result of the trial is given by this random variable.

The functional relationship between the random variable and the event is not necessarily mutually unique. Taking again the example of Fig. 1-2, the i-th event means the passing of n_i electrons: in this case, the random variable is n_i itself. But if two vacuum tubes are connected in parallel, and their overall current is investigated, the new random variable will be $n_i' = n_{i1} + n_{i2}$, and several combinations of the elementary events within the vacuum tube will pertain to a given n_i'.

2-2 DISTRIBUTIONS

Now the term "stochastic process" may be defined more precisely. By a stochastic process is meant a single parameter assembly of random variables $\xi(t)$ where parameter t covers a time-instant set continuously. For instance, the previously mentioned n_i' is a function of the time aś a parameter and is made up of a random variable defined over two event spaces. Such a stochastic process may be formed by the number of calls per unit time in a telephone exchange. A more detailed investigation of stochastic processes will be presented in Chapter 3. We will now consider the distribution of random variables, together with the parameters and characteristic types of distribution.

The distribution function $F(x)$ of a random variable ξ represents the probability of ξ having a value of less than x. Thus

$$F(x) = \mathscr{P}(\xi < x) \tag{2-4}$$

The properties of the distribution function $F(x)$ are as follows:

1. Monotonically increasing, i.e.

$$F(x_2) \geqq F(x_1), \qquad \text{if} \qquad x_2 > x_1 \tag{2-5}$$

2.
$$\lim_{x \to -\infty} F(x) = 0 \tag{2-6}$$

3.
$$\lim_{x \to \infty} F(x) = 1 \tag{2-7}$$

4. At all places, continuous from the left, i.e.

$$\lim_{x \to x_0 - 0} F(x) = F(x_0) \tag{2-8}$$

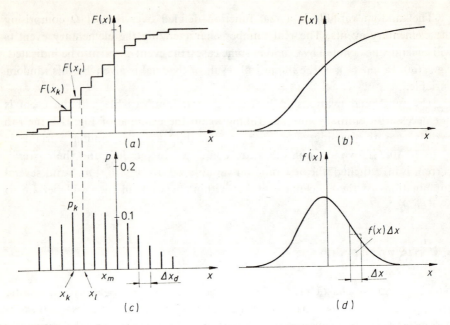

Figure 2-1 *(a)* Discrete distribution. *(b)* Continuous distribution. *(c)* Probabilities of discrete distribution *(d)* Density function of continuous distribution.

As shown in Fig. 2-1,

$$\mathscr{P}(x_k \leqq \xi < x_l) = F(x_k) - F(x_l) \tag{2-9}$$

Two basic types of a distribution function—discrete and continuous—may be distinguished. In the case of a discrete distribution, the possible members of a random variable ξ are members of a finite or infinite series. If the probability of the possible values $x_1, x_2, ..., x_i$ is $p_1, p_2, ..., p_i$, respectively, i.e.

$$P_i = \mathscr{P}(\xi = x_i) \qquad i = 1, 2, ... \tag{2-10}$$

then the distribution function is

$$F(x) = \mathscr{P}(\xi < x) = \sum_{x_i < x} p_i \tag{2-11}$$

Comparing this with Eq. (1-7) we see that if $x \to \infty$ the sum of probabilities p_i tends toward one. The distribution is said to be continuous if a function $f(x) \geqq 0$ exists for which the condition

$$F(x_2) - F(x_1) = \mathscr{P}(x_1 \leqq \xi < x_2) = \int_{x_1}^{x_2} f(x)\,dx \tag{2-12}$$

is met in any left-bounded interval (x_1, x_2).

Figures 2-1a to 2-1d show a discrete and a continuous distribution, as well as the pertaining discrete probabilities p_i and the density function $f(x)$:

$$F(x_j) - F(x_i) = p_i \qquad (2\text{-}13a)$$

and

$$\frac{dF(x)}{dx} = f(x) \qquad (2\text{-}13b)$$

where p_i can be calculated by taking the difference in two consecutive steps of the staircase distribution function, and $f(x)$ may be calculated by differentiating $F(x)$.

The density function of a discrete distribution may not be defined. However, the discrete probabilities of a series of measurements comprising a large number of data may be approximated by a continuous function. In a case where the "spectrum lines" representing the probabilities p_i are equidistant and relatively densely spaced

$$p_i = f(x)\, \Delta x_d \qquad f(x) = \frac{p_i}{\Delta x_d} \qquad (2\text{-}14)$$

where Δx_d is the distance between consecutive spectrum lines of the discrete distribution. The probability for unity Δx is just $f(x)$, which is therefore called the probability density function, or in short, the density function. Its dimension is $1/\Delta x$ where Δx may be an increment of any variable, e.g., current, voltage, resistance, time, etc.

It is clear that if step heights are decreased beyond all limits the discrete distribution will turn into a continuous distribution, and a continuous distribution may be converted into a discrete distribution by a staircase approximation.

Both conversion types will be utilized below. One of these conversions has already been applied in Sec. 1-2, where the discrete probabilities $p_i = f(n)$ of the Poisson distribution, defined only for integers, have been substituted by a continuous function because of the large values of n.

2-3 EXPECTED VALUE AND STANDARD DEVIATION

The expected value is the most important characteristic of a random variable distribution. The expected value is that number around which the algebraic mean—the so-called empirical mean—of the independent observations related to the random variable in question will fluctuate. If the random variable ξ has a discrete distribution and its values $x_1, x_2, ..., x_i$ have the probabilities $p_1, p_2, ..., p_i$ then the expected value will be the sum of the series

$$\sum_{i=1}^{\infty} x_i p_i \qquad p_i = \lim_{N \to \infty} \frac{n_i}{N} \qquad (2\text{-}15)$$

on the condition that this series is absolutely convergent. The expected value of the random variable ξ will be denoted by $M(\xi)$. This notation is derived from the initial of the word "mean", and is an operational instruction for the random variable, in parentheses.

From Eqs. (2-14) and (2-15), the expected value of a continuous distribution is

$$M(\xi) = \int_{-\infty}^{\infty} xf(x)\,dx \qquad (2\text{-}16)$$

This has to meet the condition of absolute integrability, i.e.

$$\int_{-\infty}^{\infty} |x|\,f(x)\,dx < \infty \qquad (2\text{-}17)$$

Otherwise the expected value is not defined. The formulas needed for determining the expected value are very similar to those applied to the calculation of plane figure moments referring to a specific axis. This is why $M(\xi)$ is usually called the first-order moment of the distribution (Fig. 2-2).

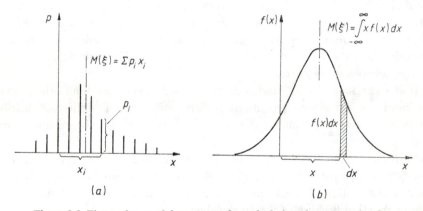

Figure 2-2 First-order partial moments for calculating the expected value. *(a)* Discrete distribution. *(b)* Continuous distribution.

A few typical properties of the expected value will now be presented:

$$M(c\xi) = cM(\xi) \qquad (2\text{-}18)$$

where c is a constant. The expected value of the sum of two or more random variables is

$$M(\xi_1 + \xi_2 + \ldots + \xi_n) = M(\xi_1) + M(\xi_2) + \ldots + M(\xi_n) \qquad (2\text{-}19)$$

It follows from Eq. (2-19) that for $\xi_1 = \xi$ and $\xi_2 = c$

$$M(\xi + c) = M(\xi) + c \qquad (2\text{-}20)$$

If ξ_1 and ξ_2 are statistically independent random variables,

$$M(\xi_1 \xi_2) = M(\xi_1) M(\xi_2) \tag{2-21}$$

The condition for independence is that the joint distribution, i.e. the probability of the simultaneous occurrence of two events pertaining to random variables ξ_1 and ξ_2 given by

$$\mathscr{P}(\xi_1 < x_1, \ \xi_2 < x_2) = G(x_1, x_2) \tag{2-22}$$

should have the density function

$$g(x_1, x_2) = f_1(x_1) f_2(x_2) \tag{2-23}$$

i.e. it should be the product of the density functions of the two variables.

The random variable fluctuates around the mean value $M(\xi)$. However, we cannot yet tell anything about the amplitude of the fluctuation. The instantaneous deviation from the expected value may be either positive or negative, thus the deviations will cancel each other out. This requires then that the measure of fluctuation should be independent of the sign. Let us therefore introduce the positive square root of the mean square value of fluctuation, the so-called standard deviation

$$D(\xi) = \sqrt{M\{[\xi - M(\xi)]^2\}} \tag{2-24}$$

This means that the standard deviation of the random variable ξ is the positive square root taken from the expected value of the squared random variable $\xi - M(\xi)$. Here, the operational instruction $D(\ )$ refers to the calculation of the dispersion of the random variable in parentheses.

The application of Eq. (2-24) is somewhat inconvenient. Let us perform the indicated operations

$$D^2(\xi) = M\{[\xi - M(\xi)]^2\} = M[\xi^2 - 2\xi M(\xi) + M^2(\xi)] \tag{2-25}$$

The factor $M(\xi)$ in the central right-hand term is constant, so by applying Eq. (2.18), we have

$$D^2(\xi) = M(\xi^2) - 2M^2(\xi) + M^2(\xi) = M(\xi^2) - M^2(\xi) \tag{2-26}$$

This means that the variance may be expressed in another way too, namely by the difference between the expected value of the squared random variable and the squared expected value. However, care should be taken when applying Eq. (2-26) as in some cases it may be much less accurate than Eq. (2-24). For example, let us assume that the expected value $M(\xi)$ has been determined with an error ε. This means that according to Eq. (2-24), we have

$$D^2(\xi) = M\{[\xi - M(\xi) - \varepsilon]^2\}$$
$$= M[\xi^2 - 2\xi M(\xi) + M^2(\xi) - 2\varepsilon\xi + 2\varepsilon M(\xi) + \varepsilon^2]$$
$$= M(\xi^2) - M^2(\xi) + \varepsilon^2 \tag{2-27}$$

The terms $-2\varepsilon\xi+2\varepsilon M(\xi)$ are canceled because $M(\xi)=M[M(\xi)]$ as the expected value of a constant number is that number itself. Let us now repeat the calculation by applying Eq. (2-26).

$$D^2(\xi) = M(\xi^2)-[M(\xi+\varepsilon)]^2$$
$$= M(\xi^2)-M^2(\xi)-2\varepsilon M(\xi)-\varepsilon^2 \qquad (2\text{-}27a)$$

where Eq. (2-20) has been utilized to calculate $M(\xi+\varepsilon)$. In the latter case, the error of $D^2(\xi)$ is substantially higher as $2\varepsilon M(\xi)\gg\varepsilon^2$ if $\varepsilon\ll M(\xi)$.

Knowledge of the density function permits the calculation of any distribution variance. According to Eq. (2-24), the variance is the expected value of the squared fluctuations, so for a discrete distribution, it follows from Eq. (2-15) that

$$D^2(\xi) = \sum_i [x_i-M(\xi)]^2 p_i \qquad (2\text{-}28)$$

and for a continuous distribution, we have from (2-17) that

$$D^2(\xi) = \int_{-\infty}^{\infty} [x-M(x)]^2 f(x)\,dx \qquad (2\text{-}29)$$

The latter relationship may be converted in a similar way to Eq. (2-26). Thus

$$D^2(\xi) = \int_{-\infty}^{\infty} x^2 f(x)\,dx - M^2(\xi) = \mu_2-\mu_1^2 \qquad (2\text{-}30)$$

where $\mu_1=M(\xi)$ is the first-order moment of the distribution. It can be seen that the integral of $x^2f(x)\,dx$ is the second-order moment of the density function taken for $x=0$, denoted by μ_2. Equation (2-30) gives the so-called central second-order moment.

Let us present two typical properties of the variance

$$D^2(a\xi+b) = M\{[a\xi+b-M(a\xi+b)]^2\}$$
$$= M\{[a\xi+b-aM(\xi)-b]^2\} = a^2 D^2(\xi) \qquad (2\text{-}31)$$

According to Eq. (2-24), the standard deviation is thus multiplied by the factor $|a|$, but is not changed by the additive term.

Another important theorem is that the variance of mutually independent random variables $\xi_1, \xi_2, ..., \xi_n$ is given by

$$D^2(\xi_1+\xi_2+...+\xi_n) = D^2(\xi_1)+D^2(\xi_2)+...+D^2(\xi_n) \qquad (2\text{-}32)$$

which may be verified as follows:

$$D^2(\xi_1+\xi_2+...+\xi_n) = M\{[\xi_1+\xi_2+...+\xi_n-M(\xi_1+\xi_2+...+\xi_n)]^2\}$$
$$= M\{[\xi_1-M(\xi_1)+\xi_2-M(\xi_2)+...+\xi_n-M(\xi_n)]^2\} \qquad (2\text{-}33)$$

In the course of squaring, we obtain the squares and double products of terms that have the form $\xi_i - M(\xi_i)$. The squared terms add up to just $D^2(\xi)$, and the double product terms cancel because

$$M\{[\xi_i - M(\xi_i)][\xi_k - M(\xi_k)]\} = M[\xi_i - M(\xi_i)]\,M[\xi_k - M(\xi_k)] = 0 \qquad (2\text{-}34)$$

The first equality follows from the independence requirement. Provisionally denoting the random variables within the brackets as η_1, η_2, it follows from Eq. (2-21) that $M(\eta_1, \eta_2) = M(\eta_1)\,M(\eta_2)$. However, the last two factors disappear separately because $M(\xi_i) = M[M(\xi_i)]$.

If the variances of independent random variables $\xi_1, \xi_2, \ldots, \xi_n$ are equal

$$D^2(\xi_1 + \xi_2 + \ldots + \xi_n) = nD^2(\xi) \qquad (2\text{-}35)$$

The presence of additive constants, i.e. the differences between the density functions of the random variables, have no influence on the end result.

As in the case of the first-order and second-order moments, the higher-order moments of distributions may also be defined by

$$\mu_n = \int_{-\infty}^{\infty} x^n f(x)\,dx \qquad (2\text{-}36)$$

The investigation of higher-order moments may yield valuable information about the density function. For example, in Fig. 2-3 symmetrical and asymmetrical density functions having all zero first-order moments and variances not differing significantly are compared. However, for the asymmetrical function $\mu_3 \neq 0$, the third-order moment is characteristic of the skewness. By comparing the

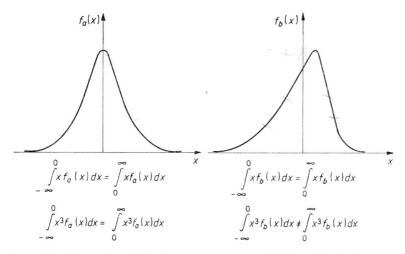

Figure 2-3 Comparison of third-order moments pertaining to distributions with zero expected values. The origin of the asymmetrical density function $f_b(x)$ is chosen for $M(\xi) = 0$.

second-order moment μ_2 and the fourth-order moment μ_4, the kurtosis of the density function may be derived. For instance, if the density function has a long tail, i.e. it is not zero even at points that are separated from the mean value by several multiples of the standard deviation, then the fourth-order moment is significantly increased.

2-4 CHARACTERISTIC FUNCTION, SUM DISTRIBUTIONS

The introduction of the characteristic function is justified partly because of the simplified calculation of higher-order moments and partly because of the investigation of so-called sum distributions. It should be mentioned in advance that the characteristic function is the Fourier transform of the density function, and this transform will be used several times. It cannot always be calculated in closed form, necessitating in these cases computer-aided methods.[18]

The characteristic function of a random variable ξ is defined for all real numbers as

$$\varphi(v) = M(e^{j\xi v}) = M(\cos \xi v) + jM(\sin \xi v) \tag{2-37}$$

In noise analysis, ξ generally depends on time; $\xi = \xi(t)$. Naturally, the characteristic function may also be defined for time independent random variables.

Taking into account the instruction for calculating the expected value, we have the following Fourier series for discrete values [see Eq. (2-15)]:

$$\varphi(v) = \sum_i e^{jx_i v} p_i = \sum_i p_i \cos x_i v + j \sum_i p_i \sin x_i v \tag{2-38}$$

In case of continuous distribution, it follows from Eq. (2-16) that

$$\varphi(v) = \int_{-\infty}^{\infty} e^{jxv} f(x)\, dx = \int_{-\infty}^{\infty} f(x) \cos vx\, dx + j \int_{-\infty}^{\infty} f(x) \sin vx\, dx \tag{2-39}$$

$\varphi(v)$ is thus seen to be truly the Fourier transform of $f(x)$. Naturally, the formula for back transforming

$$f(x) = \frac{1}{2\pi} \int_{-\infty}^{\infty} e^{-jvx} \varphi(v)\, dv \tag{2-40}$$

is also valid.

A few typical properties of the characteristic function are given, as shown, for instance, in Fig. P 2-2 (p. 54).

For $v=0$, $\varphi(v)=1$, as in this case only $\int_{-\infty}^{+\infty} f(x)\,dx$ is left from Eq. (2-39) which is equal to unity according to Eqs. (2-6) and (2-7). It may also be proven that in the case of $v \neq 0$, $|\varphi(v)| \leq 1$.

Generally $\varphi(v)$ is complex. However, if $f(x)$ is an even function, the second integral on the right-hand side of Eq. (2-39) disappears, so $\varphi(v)$ becomes real. If it is not required for $f(x)$ to be even, then

$$\varphi(-v) = \varphi^*(v) \tag{2-41}$$

This is proved by noting that $M[\cos(-\xi v)] = M(\cos \xi v)$ and $M[\sin(-\xi v)] = M[-\sin(\xi v)]$. (The asterisk denotes now and in the following the conjugated value of a complex number.)

Let us now calculate the derivatives of $\varphi(v)$ with respect to v.

$$\frac{d\varphi}{dv} = j \int_{-\infty}^{\infty} e^{jvx} x f(x)\, dx$$

$$\frac{d^2\varphi}{dv^2} = -\int_{-\infty}^{\infty} e^{jvx} x^2 f(x)\, dx$$

$$\frac{d^m\varphi}{dv^m} = j^m \int_{-\infty}^{\infty} e^{jvx} x^m f(x)\, dx \tag{2-42}$$

Substituting $v=0$, we have

$$\left.\frac{d\varphi}{dv}\right|_0 = jM(\xi) = j\mu_1$$

$$\left.\frac{d^2\varphi}{dv^2}\right|_0 = -\mu_2$$

$$\left.\frac{d^m\varphi}{dv^m}\right|_0 = j^m \mu_m \tag{2-43}$$

Thus an easy calculation of higher-order moments, which in many cases is essential for the investigation of distributions, is possible.

Utilizing the characteristic function, the sum distribution of random variables may be calculated. This is especially important because in most electronic devices or circuits, several noise sources are present simultaneously so the overall amplitude density function is made up of several components.

Let the random variable of the sum distribution be

$$\eta = \xi_1 + \xi_2 + \ldots + \xi_n \tag{2-44}$$

where $\xi_1, \xi_2, \ldots, \xi_n$ are mutually independent. Then

$$\varphi_\eta(v) = M(e^{jv\eta}) = M(e^{jv\xi_1}) M(e^{jv\xi_2}) \ldots M(e^{jv\xi_n}) \tag{2-45}$$

as the overall expected value is the product of the individual expected values because of the condition of mutual independence. This means that the Fourier transform of the sum distribution density function is the product of the Fourier transforms corresponding to the individual density functions.

Let us, for the time being, restrict our considerations to the two-term sum

$$\eta = \xi_1 + \xi_2 \tag{2-46}$$

and let the density functions of ξ_1, ξ_2 and η be $f_1(x_1)$, $f_2(x_2)$ and $f(y)$, respectively·
Using the Fourier transform to transform back the product of the characteristic functions

$$\varphi_\eta(v) = \varphi_1(v)\varphi_2(v) \tag{2-47}$$

we have the well-known convolution integral

$$f(y) = \frac{1}{2\pi} \int_{-\infty}^{\infty} e^{-jvy}\varphi_1(v)\varphi_2(v)\,dv = \int_{-\infty}^{\infty} f_1(x_1)f_2(y-x_1)\,dx_1 \tag{2-48}$$

This can be derived using another reasoning too. The joint probability of independent events is given by the product of their probabilities, so $f_1(x_1)$ and $f_2(x_2)$ are to be multiplied, but the condition of $x_2 = y - x$, for any value of y, should also be met. Covering all possible values of x_1, the density function of y is calculated by the addition of the partial products, i.e. by integration in the case of a continuous density function. A few examples of the convolution of typical density functions representing electronic noise are given by Arend.[1]

A simple example will now be given to explain the density function of a sum distribution. Let us consider the throwing of two dice, and suppose that our interest in the outcome extends to our knowing the sum of the numbers appearing on the dice. What are the probabilities of the possible sum values? The probability of any number appearing when throwing a single die is 1/6. The probability that a six will appear with both dice is 1/36 as we have independent events. The probability of obtaining 2 and 12 as a sum will be only 1/36, but, for instance, the number five may be cast in four different ways: i.e. 1+4, 2+3, 3+2, 4+1. Expressing Eq. (2-48) for a discrete distribution, in our case we have

$$p(5) = \sum_{i=1}^{4} p_i p_{(5-i)} = 4 \times \frac{1}{6} \times \frac{1}{6} = \frac{1}{9} \tag{2-49}$$

Figure 2.4a shows the discrete distributions for throwing dice, and the corresponding continuous, so-called uniform distribution is shown in Fig. 2.4b. Given two random variables having uniform distributions, the density function of their sum is the convolution of two rectangular density functions: this convolution is triangular.

Before considering the central limit theorem, the reader is asked to calculate the sum density function of three random variables having equal uniform distributions (e.g., the tossing of three dice). Furthermore, by increasing the number of members giving the sum, how will the overall density function be modified?

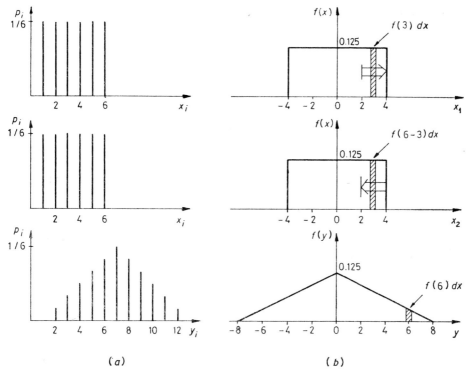

(a)   (b)

Figure 2-4 Discrete and continuous sum distributions. *(a)* Probabilities of dice casting and sum distribution. *(b)* Convolution of uniform distributions.

2-5 THE CENTRAL LIMIT THEOREM

The noise appearing in devices and circuits may frequently be considered as a superposition of many random phenomena with equal distributions. An important question is: can the density function of the sum distribution be determined in this case, especially when the number of components is high and unknown?[6, 16] According to the central limit theorem of mathematical statistics, the distribution of a sum made up of a large number of independent random variables follows normal (gaussian) distribution.

Let $\xi_1, \xi_2, ..., \xi_i, ..., \xi_n$ be mutually independent random variables with zero expected values and σ_i standard deviations; furthermore $\mu_3 = 0$ skewness.* Let the sum random variable be defined by

$$\eta = \sum_{i=1}^{n} \xi_i \tag{2-50}$$

---

* This requirement simplifies the derivation. However, the central limit theorem may also be derived with more general starting-point conditions.

According to Eq. (2-19), this has an expected value of

$$M(\eta) = \sum_{i=1}^{n} M(\xi_i) = nM(\xi_i) = 0 \tag{2-51}$$

and a variance, according to Eq. (2-35), of

$$\sigma^2 = D^2(\eta) = \sum_{i=1}^{n} D^2(\xi_i) = nD^2(\xi_i) = n\sigma_i^2 \tag{2-52}$$

Furthermore

$$\mu_4(\eta) = M(\eta^4) \qquad \mu_4(\xi) = M(\xi^4) \tag{2-53}$$

Let us normalize the sum random variable η to the standard deviation σ:

$$z = \frac{\eta}{\sigma} = \frac{\xi_1 + \xi_2 + \ldots + \xi_n}{\sigma} \tag{2-54}$$

The variances and fourth-order moments of the basic distributions are normalized to σ^2 and σ^4, respectively. Thus

$$\frac{D^2(\xi_i)}{\sigma^2} = \frac{\sigma_i^2}{\sigma^2} = \frac{1}{n} \tag{2-55}$$

and

$$\frac{\mu_4(\xi_i)}{\sigma^4} = \frac{\mu_4(\xi_i)}{n^2 \sigma_i^4} \tag{2-56}$$

Using Eq. (2-45), let us express the characteristic function of the sum distribution as

$$\varphi(v) = M(e^{jzv}) = \prod_{i=1}^{n} M(e^{j(v/\sigma)\xi_i}) = \left[\varphi_i\left(\frac{v}{\sigma}\right)\right]^n \tag{2-57}$$

The function $\varphi_i\left(\dfrac{v}{\sigma}\right)$ may be expanded to

$$\varphi_i\left(\frac{v}{\sigma}\right) = M\left[1 + j\left(\frac{v}{\sigma}\right)\xi_i - \frac{1}{2!}\left(\frac{v}{\sigma}\right)^2 \xi_i^2 - \frac{j}{3!}\left(\frac{v}{\sigma}\right)^3 \xi_i^3 + \ldots\right]$$

$$= 1 + j\left(\frac{v}{\sigma}\right)M(\xi_i) - \frac{1}{2!}\left(\frac{v}{\sigma}\right)^2 D^2(\xi_i) - \frac{j}{3!}\left(\frac{v}{\sigma}\right)^3 \mu_3(\xi_i) + \ldots \tag{2-58}$$

Taking into account that $M(\xi_i)=0$ and $\mu_3(\xi_i)=0$, and utilizing Eqs. (2-55) and (2-56), we have

$$\varphi_i\left(\frac{v}{\sigma}\right) = 1 - \frac{1}{2}\frac{v^2}{n} + \frac{1}{4!}\frac{v^4}{n^2} - + \ldots \tag{2-59}$$

In Eq. (2-57), we have the product of functions φ_i. Let us therefore take the logarithm of φ_i and apply the approximation $\ln(1+y) \cong y$. If $n \to \infty$ and $y \to 0$

$$\ln \varphi_i\left(\frac{v}{\sigma}\right) \cong \ln\left(1 - \frac{v^2}{2n}\right) \cong -\frac{v^2}{2n} \tag{2-60}$$

Then, according to Eq. (2-57), we have

$$\ln \varphi(v) = n \ln \varphi_i \left(\frac{v}{\sigma}\right) = -\frac{v^2}{2} \tag{2-61}$$

and so

$$\varphi_v = e^{-v^2/2} \tag{2-62}$$

This is the Fourier transform of the sum distribution density function. Applying the inverse Fourier transform, we have

$$f(x) = \frac{1}{2\pi} \int_{-\infty}^{\infty} e^{-v^2/2} \cos vx \, dv = \frac{1}{\sqrt{2\pi}} e^{-x^2/2} \tag{2-63}$$

(see Appendix B). Here the following simplification has been applied: $e^{-v^2/2}$ being an even function, the integral of $e^{-v^2/2} \sin vx$, taken between plus and minus infinity, is zero.

Equation (2-63) is the density function of an important and frequently occurring distribution called normal or gaussian distribution. In this form, it applies to a distribution with a mean value of $M(z)=0$ and, because of the normalization according to Eq. (2-54), with unity standard deviation. If $M(z)=\mu_1 \neq 0$ and $D^2(z)=\sigma^2 \neq 1$, then in general form,

$$\boxed{f(x) = \frac{1}{\sigma \sqrt{2\pi}} e^{-(x-\mu_1)^2/2\sigma^2}} \tag{2-64}$$

It is worthwhile mentioning that for $n \to \infty$, i.e. when the density function of a sum made up of many random variables with equal distributions has to be determined, the result will always be a normal distribution; the only requirement to be met is that the basic distributions should not be skew.

The random variable ξ in the fluctuation phenomena is a function of time t. In this case, the density function gives the probability of individual instantaneous values. For instance, if an infinite number of equal-amplitude, incoherent sine waves is superimposed (see Fig. 1.4b for the peaked amplitude density function), we again have instantaneous values with normal distribution.* It is therefore not surprising that a stochastic signal occupies a broad frequency range. It should be stressed, however, that a "white" power spectrum in itself does not necessarily imply instantaneous values of normal distribution.

We have seen that meeting a single moderate requirement is sufficient always to arrive, in a limiting case, at the normal distribution. It follows from this fact

---

* This may be derived, for instance, from Eq. (30) in Chapter 7 of the book by Miller,[10] after expansion and neglections.

that the distribution of a sum made up of two random variables with equal expected values and normal distributions is also normal. This fact will be utilized in calculating the instantaneous values and powers of noises originating from different sources.

2-6 NORMAL DISTRIBUTION

Normal distribution is of special importance, so a separate Section has been assigned to present its properties. Its density function is given, in the general case, by Eq. (2-64):

$$f(x) = \frac{1}{\sigma \sqrt{2\pi}} e^{-(x-\mu_1)^2/2\sigma^2}$$

and the distribution function, according to Eq. (2-9), by

$$F(x) = \int_{-\infty}^{x} f(u)\, du = \Phi\left(\frac{x-\mu_1}{\sigma}\right) \tag{2-65}$$

The latter may not be expressed in closed form, but is tabulated in many books (see, for instance, Jahnke and Emde[9]) with parameters $\mu_1=0$ and $\sigma=1$. For $\mu_1=0$, the first few terms of its Taylor series are given by

$$\Phi\left(\frac{x}{\sigma}\right) = \frac{1}{2} + \frac{1}{\sqrt{2\pi}}\left[\left(\frac{x}{\sigma}\right) - \frac{1}{3!}\left(\frac{x}{\sigma}\right)^3 + \frac{1\times3}{5!}\left(\frac{x}{\sigma}\right)^5 - + \cdots\right] \tag{2-66}$$

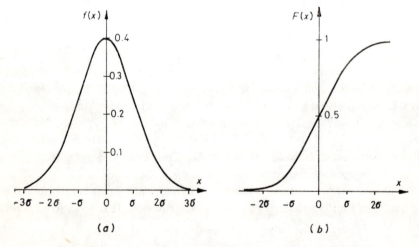

(a)   (b)

Figure 2-5 Normal (gaussian) distribution. (a) Density function.
(b) Distribution function.

The density function and the distribution function are shown in Fig. 2-5. The question frequently asked is what is the probability of the instantaneous value of a signal with normal distribution falling between the limits $x = \pm k\sigma$? These probabilities for a few normalized values are shown in Table 2-1.

Table 2-1 Tabulated values of the error function

$\dfrac{x}{\sigma}$	$\Phi\left(\dfrac{x}{\sigma}\right) - \Phi\left(-\dfrac{x}{\sigma}\right)$
1	0.683
2	0.954
3	0.9972
4	0.99994

The expected value, i.e. the first-order moment of a normal distribution symmetrical with respect to $x = 0$, is zero. However, let us calculate the expected value of the half-distribution pertaining to positive x values according to Eq. (2-16)

$$M(\xi) = \int_0^\infty xf(x)\,dx \tag{2-67}$$

$$= \frac{1}{\sigma\sqrt{2\pi}} \int_0^\infty xe^{-x^2/2\sigma^2}\,dx = \frac{\sigma}{\sqrt{2\pi}} \cong 0.4\sigma \tag{2-68}$$

$$x \geq 0$$

This will be characteristic for a half-wave detection of a normal distribution signal. For full-wave detection, we have

$$M(|\xi|) = 2 \int_0^\infty xf(x)\,dx = \sqrt{\frac{2}{\pi}}\,\sigma \cong 0.8\sigma \tag{2-69}$$

Now let us calculate the higher-order moments by multiple differentiation of the characteristic function. Let $\varphi(v)$ now be the Fourier transform of a normal distribution with zero mean value and arbitrary standard deviation [see Appendix B]. Thus

$$\varphi(v) = \frac{1}{\sigma\sqrt{2\pi}} \int_{-\infty}^\infty \cos vx\, e^{-x^2/2\sigma^2}\,dx = e^{-\sigma^2 v^2/2} \tag{2-70}$$

The derivatives and the moments are

$$\frac{d\varphi}{dv} = -\sigma^2 v e^{-\sigma^2 v^2/2}\big|_{v=0} = 0 = j\mu_1 \tag{2-71}$$

$$\frac{d^2\varphi}{dv^2} = \sigma^2 e^{-\sigma^2 v^2/2}(\sigma^2 v^2 - 1)_{v=0} = -\sigma^2 = -\mu_2 \tag{2-72}$$

$$\frac{d^3\varphi}{dv^3} = \sigma^4 v e^{-\sigma^2 v^2/2}(3 - \sigma^2 v^2)_{v=0} = 0 = -j\mu_3 \tag{2-73}$$

$$\frac{d^4\varphi}{dv^4} = \sigma^4 e^{-\sigma^2 v^2/2}(3 - 6\sigma^2 v^2 + \sigma^4 v^4)_{v=0} = 3\sigma^4 = \mu_4 \tag{2-74}$$

An important property of a normal distribution is that

$$\frac{\mu_4}{\mu_2^2} = 3 \tag{2-75}$$

Naturally, Eqs. (2-19) and (2-32) are valid as before if $\xi_1, \xi_2, ..., \xi_n$ are independent random variables with normal distribution. Since $\eta = \xi_1 + \xi_2 + ... + \xi_n$,

$$M(\eta) = M(\xi_1) + M(\xi_2) + ... + M(\xi_n)$$
$$D^2(\eta) = D^2(\xi_1) + D^2(\xi_2) + ... + D^2(\xi_n)$$

In this case, according to Sec. 2-4, we may be sure that η also has a normal distribution.

2-7 DISCRETE DISTRIBUTIONS

In probability theory, there are many types of discrete distributions. We are interested in only two of these: the previously mentioned binomial distribution and Poisson distribution [see Eqs. (1-1) and (1-4)].

Let us consider two sets of events that may have only two, mutually exclusive outcomes. This is called a simple alternative. An example of this is the emission or nonemission of an electron from the vacuum-tube cathode. Another example is related to two electrodes with the possibility of either of them being the target for an emitted electron. The noise originating from the random distribution of current fluctuation may also be treated using this model.

Let $\mathscr{P}(A) = p_1$ be the probability of the event in question, i.e. that electrode A is hit by the electron. The probability of the event not occurring is $p_2 = 1 - p_1$. Thus in the first case the value of the random variable is $\xi = x_1 = 1$, and in the second case, $\xi = x_2 = 0$. The expected value of ξ is then

$$M(\xi) = \sum_{i=1}^{2} p_i x_i = 1 \cdot p_1 + 0 \cdot p_2 = p_1 \tag{2-76}$$

and its variance is

$$D^2(\xi) = M(\xi^2) - M^2(\xi) = \sum_{i=1}^{2} p_i x_i^2 - M^2(\xi)$$

$$= 1 \cdot p_1 + 0 \cdot p_2 - p_1^2 = p_1(1 - p_1) = p_1 p_2 \qquad (2\text{-}77)$$

Let us repeat the above experiment N times. What is the probability of the event occurring just n times, and not occurring $N-n$ times? If the single experiments are independent from each other, then the overall probability is given by the product of the single probabilities, i.e.

$$\prod_n p_1 = p_1^n \qquad \text{and} \qquad \prod_{N-n} p_2 = p_2^{N-n} \qquad (2\text{-}78)$$

However, as there is no restriction regarding the sequence of outcomes, n events from a total of N may occur in $\binom{N}{n}$ different ways. Thus

$$\boxed{\mathscr{P}(\eta = n) = \binom{N}{n} p_1^n p_2^{N-n}} \qquad (2\text{-}79)$$

which are probabilities according to the binomial distribution. Here the random variable η may be defined according to

$$\eta = \xi_1 + \xi_2 + \ldots + \xi_N \qquad (2\text{-}80)$$

where $\xi_i = 1$ or 0.

Because of the requirement for independent alternatives, we have from Eq. (2-19)

$$M(\eta) = NM(\xi) = Np_1 \qquad (2\text{-}81)$$

and from Eq. (2-32)

$$D^2(\eta) = ND^2(\xi) = Np_1 p_2 \qquad (2\text{-}82)$$

If $p_1 \leq 1$, $p_2 = 1 - p_1 \cong 1$ and

$$D^2(\eta) = M(\eta) \qquad (2\text{-}83)$$

If, on the other hand, $p_1 = 1/2$, $p_2 = 1/2$ and

$$D^2(\eta) = \frac{M(\eta)}{2} = \frac{N}{4} \qquad (2\text{-}84)$$

Figure 2-6 shows the probabilities according to the binomial distribution for $N = 10$ and $p_1 = 0.1$, as well as for $N = 10$ and $p_1 = 0.5$.

If single experiments have several outcomes instead of simple alternatives, their probability is given by the polynomial distribution. However, as in actually utilized electronic devices the number of independent electron- (hole-) collecting electrodes is normally not more than two, the polynomial distribution is seldom used.

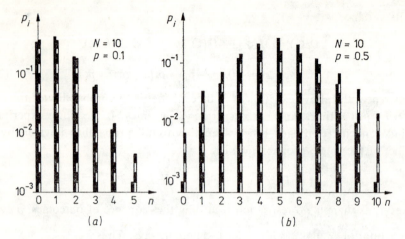

Figure 2-6 Ten-term binomial distribution. *(a)* $p=0.1$. *(b)* $p=0.5$.
(For comparison, the dashed columns show the Poisson distributions for $\lambda=1$ and $\lambda=5$.
The logarithmic scale of p is compressed in the vertical direction.)

Equations (1-2) to (1-4) have shown that if $N\to\infty$ and $p\to 0$ so that $Np=$ $=\lambda>0$, then the binomial distribution goes to the Poisson distribution given by

$$\mathscr{P}(\eta = n) = \frac{\lambda^n}{n!}\,e^{-\lambda} = f(n) \tag{2-85}$$

Thus if the above conditions are met, then the difference between the two distributions will be arbitrarily small for all n values. Figure 2-6 shows also the probabilities according to the Poisson distribution. It is noted that for $N=10$, $p=0.1$ the approximation is fair, but for $N=10$, $p=0.5$ the approximation is not applicable, especially at the extremes of the distribution. Thus if the analysis of a specific problem leads to a binomial distribution then, before a Poisson approximation is carried out, its justification should be determined. [For instance, a simple calculation and comparison can be made for $\mathscr{P}(\eta=0)$.] As an example, the intervals Δt shown in Fig. 1-1 may have been refined arbitrarily, so the condition $p\to 0$ could be met. On the other hand, the probability of current distribution between two electrodes is given $(0<p_1<1$ and $0<p_2=$ $=1-p_1<1$, respectively), so the refinement of the time intervals, or the meeting of the condition $N\to\infty$ in itself will not allow the Poisson approximation.

The moments of the Poisson distribution may be calculated from the characteristic function

$$\varphi(v) = \sum_{n=0}^{\infty} e^{jvn}\frac{\lambda^n}{n!}e^{-\lambda} = e^{-\lambda}\sum_{n=0}^{\infty}\frac{(\lambda e^{jv})^n}{n!} = e^{-\lambda}e^{\lambda e^{jv}} = e^{\lambda(e^{jv}-1)} \tag{2-86}$$

where $e^x = \sum\limits_{0}^{\infty} x^n/n!$ has been used. The first derivative is

$$\frac{d\varphi}{dv}\Big|_{v=0} = j\lambda e^{jv} e^{\lambda(e^{jv}-1)}\big|_{v=0} = j\lambda = jM(\eta) \tag{2-87}$$

and yields j times the expected value. The second derivative is

$$\frac{d^2\varphi}{dv^2} = -\lambda^2 e^{2jv} e^{\lambda(e^{jv}-1)} - \lambda e^{jv} e^{\lambda(e^{jv}-1)}\big|_{v=0} = -\lambda^2 - \lambda = -\mu_2 \tag{2-88}$$

Utilizing Eq. (2-30), we have

$$D^2(\eta) = \mu_2 - \mu_1^2 = \lambda \tag{2-89}$$

The expected value and the variance have already been utilized in Chapter 1 to calculate the shot noise.

In these cases, $M(\eta)>0$ as the Poisson distribution is defined only for positive integers. The probabilities of the distribution are shown in Fig. 2-7. It is seen that the "envelope" which has only an illustrative role for discrete random variables is not symmetrical with respect to $M(\eta)$, so generally $\mu_3(\eta)\neq 0$. In spite of this, the normal distribution may be used as a good approximation for high enough values of n and λ, if n is near λ (i.e. not for arbitrarily high n values).

For large n values, the Stirling formula*

$$n! = n^n e^{-n} \sqrt{2\pi n} \tag{2-90}$$

may be applied. Let us substitute this into Eq. (2-85) and let us take the natural logarithm of Eq. (2-85)

$$\ln \mathscr{P} = -n \ln \frac{n}{\lambda} + n - \lambda - \frac{1}{2} \ln n - \frac{1}{2} \ln 2\pi \tag{2-91}$$

Let us make use of the identity

$$n \equiv \lambda \left(1 + \frac{n-\lambda}{\lambda}\right) = \lambda \left(1 + \frac{\Delta n}{\lambda}\right) = \lambda(1+\alpha) \tag{2-92}$$

and take into account that if $n \to \lambda$, then $\alpha \ll 1$ and

$$\ln(1+\alpha) = \alpha - \frac{1}{2}\alpha^2 + \dots \tag{2-93}$$

Equation (2-91) may thus be rewritten as

$$\ln \mathscr{P} = -\lambda(1+\alpha)\left(\alpha - \frac{\alpha^2}{2} + \dots\right) + \lambda\alpha - \frac{1}{2}\ln\lambda - \frac{1}{2}\left(\alpha - \frac{\alpha^2}{2} + \dots\right) - \frac{1}{2}\ln 2\pi$$

$$= -\frac{1}{2}\lambda\alpha^2 + \frac{1}{2}\lambda\alpha^3 - \frac{1}{2}\alpha + \frac{1}{4}\alpha^2 - \frac{1}{2}\ln 2\pi\lambda + -\dots \tag{2-94}$$

* A more accurate approximation of $n!$ is available.[17]

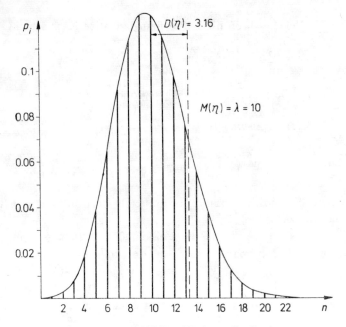

Figure 2-7 Probabilities of Poisson distribution.

or

$$\ln \mathscr{P} = -\frac{1}{2}\frac{\varDelta n^2}{\lambda} + \frac{1}{2}\frac{\varDelta n^2}{\lambda^2}\left(\varDelta n + \frac{1}{2}\right) - \frac{1}{2}\frac{\varDelta n}{\lambda} - \frac{1}{2}\ln 2\pi\lambda + \dots \qquad (2\text{-}95)$$

As $\varDelta n \ll \lambda$ but $\varDelta n = 0, 1, 2, \dots$, the two middle terms may be neglected compared with the two outer terms. Thus

$$\mathscr{P}(\eta = n) = \frac{1}{\sqrt{2\pi\lambda}}\,e^{-\varDelta n^2/2\lambda} = \frac{1}{D(\eta)\sqrt{2\pi}}\,e^{-\varDelta n^2/2D^2(\eta)} \qquad (2\text{-}96)$$

because according to Eq. (2-89) $D(\eta) = \sqrt{\lambda}$.

Considering continuous distribution, we arrive at the density function of the gaussian distribution. It may thus be concluded that if $D(\eta)$ is sufficiently small compared with $M(\eta)$, i.e. the relative standard deviation is

$$\frac{D(\eta)}{M(\eta)} = \frac{\sqrt{\lambda}}{\lambda} = \frac{1}{\sqrt{\lambda}} \ll 1 \qquad (2\text{-}97)$$

the Poisson distribution may be approximated by the normal distribution, at least in the vicinity of $M(\eta)$.* The condition for this is a sufficiently high average current and thus a sufficiently high λ. With the exception of rare cases, the shot noise of electronic devices will therefore be regarded as having normal distribution.

* The relative error may be high far from $M(\eta)$. (See Hubbard.[8])

The binomial distribution that may not be approximated by Poisson distribution has still to be investigated $(N \to \infty$, but p does not go to zero). This is really a sum distribution made up of alternative event pairs; for large numbers, the validity of the central limit theorem may be expected. However, a condition of this—the symmetrical basic distribution—is not generally met, as may be seen from Eq. (2-76). Therefore, the distribution in the vicinity of $n = Np_1 = M(\eta)$ will be investigated separately. Here

$$\mathscr{P}(\eta = Np_1) = \mathscr{P}_0 = \binom{N}{Np_1} p_1^{Np_1} p_2^{N-Np_1} = \frac{N! \, p_2^N}{(Np_1)!(N-Np_1)!} \left(\frac{p_1}{1-p_1}\right)^{Np_1} \quad (2\text{-}98)$$

Let us determine the relative probability of $\eta = Np_1 + k \ (k \ll Np_1)$. For this, $(Np_1)!$ has to be increased to $(Np_1 + k)!$ which means evidently a multiplication by $(Np_1 + 1)(Np_1 + 2)\ldots(Np_1 + k)$. Similarly, $(N - Np_1)!$ has to be divided by $(N - Np_1)(N - Np_1 - 1)\ldots[N - Np_1 - (k-1)]$. Thus

$$\frac{\mathscr{P}_k}{\mathscr{P}_0} = \frac{\mathscr{P}(\eta = Np_1 + k)}{\mathscr{P}(\eta = Np_1)}$$

$$= \frac{(N - Np_1)(N - Np_1 - 1)\ldots[N - Np_1 - (k-1)]}{(Np_1 + 1)(Np_1 + 2) \ldots (Np_1 + k)} \left(\frac{p_1}{1-p_1}\right)^k \quad (2\text{-}99)$$

A common factor in the numerator is $(N - Np_1)^k$, and in the denominator is $(Np_1)^k$. Their quotient multiplied by the last factor in Eq. (2-99) gives unity. Let us now take the natural logarithm of the remaining equation below

$$\frac{\mathscr{P}_k}{\mathscr{P}_0} = \frac{[1 - 1/(N - Np_1)][1 - 2/(N - Np_1)] \ldots [1 - (k-1)/(N - Np_1)]}{(1 + 1/Np_1)(1 + 2/Np_1)\ldots(1 + k/Np_1)}$$

Its natural logarithm is given by

$$\ln \mathscr{P}_k = \ln \mathscr{P}_0 + \sum_{i=0}^{k-1} \ln \left(1 - \frac{i}{N - Np_1}\right) - \sum_{i=1}^{k} \ln \left(1 + \frac{i}{Np_1}\right) \quad (2\text{-}100)$$

Because of the requirement that $k \ll Np_1$, the approximation $\ln(1+x) \cong x$ may be used. Thus

$$\ln \mathscr{P}_k = \ln \mathscr{P}_0 - \sum_{i=0}^{k-1} \frac{i}{N - Np_1} - \sum_{i=1}^{k} \frac{i}{Np_1} \quad (2\text{-}101)$$

Both summations will result in sums of algebraic series. Thus

$$\ln \mathscr{P}_k = \ln \mathscr{P}_0 - \frac{k}{2N}\left(\frac{k-1}{1-p_1} + \frac{k+1}{p_1}\right) = \ln \mathscr{P}_0 - \frac{k(k+1-2p_1)}{2Np_1(1-p_1)} \quad (2\text{-}102)$$

Note that according to Eq. (2-82), $2Np_1(1-p_1) = 2D^2(\eta)$ and the numerator has an approximate value of k^2. For $p_1 = 0.5$, i.e. for symmetrical distribution, this value is accurate. So

$$\mathscr{P}_k = \mathscr{P}_0 e^{-k^2/2D^2(\eta)} \quad (2\text{-}103)$$

$\mathscr{P}_0$ may be calculated from Eq. (2-98) by utilizing the Stirling formula. Thus

$$N! = N^N e^{-N} \sqrt{2\pi N} \tag{2-104}$$

$$(Np_1)! = N^{Np_1} p_1^{Np_1} e^{-Np_1} \sqrt{2\pi Np_1} \tag{2-105}$$

$$(N - Np_1)! = N^{N(1-p_1)} (1-p_1)^{N(1-p_1)} e^{-N(1-p_1)} \sqrt{2\pi N(1-p_1)} \tag{2-106}$$

After simplification, we have

$$\mathscr{P}_0 = \frac{1}{D(\eta) \sqrt{2\pi}} \tag{2-107}$$

which is the expected coefficient of the well-approximating gaussian distribution.

It may be concluded from this derivation that for sufficiently high $M(\eta)$, which is nearly always the case for charge carriers in electronic devices, the binomial distribution may be substituted by a normal distribution.

2-8 CONTINUOUS DISTRIBUTIONS

The processes within electronic devices are in connection with elementary charge carriers and should strictly speaking be characterized by discrete distributions only. However, due to the large number of charge carriers, in many cases the approximation given in Eq. (2-14) may be applied with the change over to continuous distribution. In this category, the uniform distribution is the most simple.

The continuous random variable ξ has a uniform distribution if in the interval (a, b) the density function is

$$f(x) = \frac{1}{a-b} \qquad a < x < b \tag{2-108}$$

and outside this interval is $f(x)=0$. It should again be borne in mind that the density function describing the physical process is generally not dimensionless. In Eq. (2-108), x may have any physical interpretation. In our case, it is voltage or current, so $[f(x)]=V^{-1}$ or A^{-1}.

Let us confine ourselves to the uniform distribution with zero expected value as shown in Fig. 2-4b, with limits $-a$ and $+a$. The density function is then

$$f(x) = \frac{1}{2a} \qquad -a < x < a \tag{2-109}$$

which meets the condition of Eq. (2-12)

$$\int_{-\infty}^{\infty} f(x)\, dx = 1$$

According to Eq. (2-30), the variance is given by

$$D^2(\xi) = \int_{-a}^{a} x^2 \frac{1}{2a} \, dx = \frac{x^3}{6a} \bigg|_{-a}^{a} = \frac{a^2}{3} \tag{2-110}$$

and the standard deviation by

$$D(\xi) = \frac{a}{\sqrt{3}} \tag{2-111}$$

Note that the dimension of the standard deviation is in this case again equal to the dimension of the random variable.

The fourth-order moment is

$$\mu_4 = \int_{-a}^{a} x^4 \frac{1}{2a} \, dx = \frac{a^4}{5} \tag{2-112}$$

According to Sec. 2-3, the kurtosis of the uniform distribution will be characterized by

$$\frac{\mu_4}{\mu_2^2} = \frac{\mu_4}{D^4(\xi)} = \frac{9}{5} < 3 \tag{2-113}$$

The next three types of continuous distribution will be derived from the normal distribution (χ^2, average absolute deviation, Rayleigh distribution). All three distributions are mainly applied to methods of measuring stochastic signals.

For instance, in the case of quadratic detection, the instantaneous value of the voltage or current to be measured is squared (by using a nonlinear network with infinite speed and zero time constant), and the signal thus generated is averaged or integrated. Thus the distribution of squared instantaneous values having normal distribution should be known, as should the k-term sum distribution, derived from the former distribution (if k-independent values are taken during averaging).

Let us consider as a starting point, a normal distribution of zero expected value and standard deviation σ. The probability of $\xi^2 < y$ is given by

$$F(y) = \mathscr{P}(\xi^2 < y) = \mathscr{P}(-\sqrt{y} < \xi < \sqrt{y}) \tag{2-114}$$

As the distribution of ξ is normal and symmetrical with respect to $x=0$, we have

$$F(y) = \frac{2}{\sigma \sqrt{2\pi}} \int_{0}^{\sqrt{y}} e^{-u^2/2\sigma^2} \, du = 2 \left[\Phi\left(\frac{\sqrt{y}}{\sigma}\right) - \Phi(0) \right] \tag{2-115}$$

The density function is given by

$$f(y) = \frac{dF(y)}{dy} = \frac{d}{d\sqrt{y}} 2\Phi\left(\frac{\sqrt{y}}{\sigma}\right) \frac{d\sqrt{y}}{dy} = \frac{e^{-y/2\sigma^2}}{\sigma \sqrt{2\pi y}} \tag{2-116}$$

This is shown in Fig. 2-8, and may only be defined for $y>0$. This is called a chi-square distribution with one degree of freedom (χ^2 distribution).

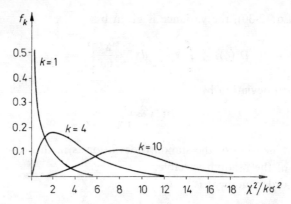

Figure 2-8 χ^2 distributions for squared sums having different number of terms.

Consider now a random variable that is not the square of a single instantaneous value ξ but the sum of squared random variables $\xi_1, \xi_2, ..., \xi_k$ having the same normal distribution

$$\chi^2 = \xi_1^2 + \xi_2^2 + ... + \xi_k^2 \qquad (2\text{-}117)$$

We then have a χ^2 distribution of k degrees of freedom, having a density function

$$\boxed{f_k(y) = \frac{y^{(k/2)-1} e^{-y/2\sigma^2}}{2^{k/2} \Gamma(k/2)\sigma^k} = \frac{x^{k-2} e^{-x^2/2\sigma^2}}{2^{k/2} \Gamma(k/2)\sigma^k} = f_k(x^2)} \qquad (2\text{-}118)$$

where f_k is given both as a function of the variable x pertaining to the original normal distribution and as a function of the variable $y = x^2$ of the χ^2 distribution. The relation may be derived either by convolution[11] or by raising to the appropriate power the characteristic function of Eq. (2-116) [see Eq. (2-45)] and transforming it back. $\Gamma(v)$ is the generalization of function $v!$ for an arbitrary nonintegral number, defined by the integral[3]

$$\Gamma(v) = \int_0^\infty x^{v-1} e^{-x} \, dx$$

In the form given in Eq. (2-118), $f_k(x)$ and $f_k(y)$ are suitable for describing a physical variable of any dimension; naturally $[x] = [\sigma]$.

Let us calculate the expected value and standard deviation of the χ^2 distribution. As $\xi_1, \xi_2, ..., \xi_k$ are independent, utilizing Eqs. (2-19) and (2-32) we have

$$M(\chi^2) = M(\xi_1^2) + M(\xi_2^2) + ... + M(\xi_k^2) = kM(\xi^2) \qquad (2\text{-}119)$$

$$D^2(\chi^2) = kD^2(\xi^2) = k[M(\xi^4) - M^2(\xi^2)] \qquad (2\text{-}120)$$

At the right-hand sides of these equations there are the second- and fourth-order moments of the basic distribution. Due to the requirement that ξ should have a normal distribution

$$D^2(\chi^2) = k(3\sigma^4 - \sigma^4) = 2k\sigma^4 \qquad (2\text{-}121)$$

Furthermore

$$M(\chi^2) = k\sigma^2 \qquad (2\text{-}122)$$

The relative standard deviation of the χ^2 distribution is

$$\frac{D(\chi^2)}{M(\chi^2)} = \sqrt{\frac{2}{k}} \qquad (2\text{-}123)$$

Figure 2-8 shows functions pertaining to different k parameters. If k is a large number, the χ^2 distribution also goes to the normal distribution as do the sum distributions in general.

In the case of full-wave, piecewise linear detection, the distribution of the detected output signal follows the distribution of $|\xi|$. If before detection ξ had a normal distribution, the density function of $|\xi|$ will have the form shown in Fig. 2-9

$$f(x) = \frac{1}{\sigma}\sqrt{\frac{2}{\pi}}\,e^{-x^2/2\sigma^2} \qquad 0 < x < \infty \qquad (2\text{-}124)$$

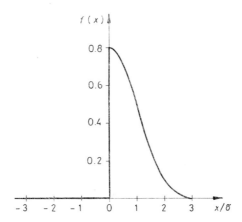

Figure 2-9 Distribution of absolute value

The expected value [see Eq. (2-69)] is

$$M(|\xi|) = \frac{1}{\sigma}\sqrt{\frac{2}{\pi}}\int_0^\infty x e^{-x^2/2\sigma^2}\,dx = \sqrt{\frac{2}{\pi}}\,\sigma$$

and the variance is

$$D^2(|\xi|) = \mu_2 - \mu_1^2 = \frac{1}{\sigma}\sqrt{\frac{2}{\pi}}\int_0^\infty x^2 e^{-x^2/2\sigma^2}\,dx - \frac{2}{\pi}\sigma^2 = \left(1 - \frac{2}{\pi}\right)\sigma^2 \qquad (2\text{-}125)$$

Naturally, during the course of measurement, the output signal of a linear detector is also averaged, so the distribution

$$\eta = |\xi_1| + |\xi_2| + \ldots + |\xi_k| \qquad (2\text{-}126)$$

of the k-term sum formed by the absolute values of the independent random variables having normal distribution is again needed. For a small number of terms, the density function[7] is

$$f(\eta) = k^{3/2} 2^{-(k+1)/2} \pi^{(k-1)/2} \sum_{i=1}^{k-1} \binom{k}{i} e^{-\eta^2 k^3/8 i(k-i)} G_{i-1}\left(\frac{k\eta}{2}\right) G_{k-i-1}\left(\frac{k\eta}{2}\right) \qquad (2\text{-}127)$$

where $G_i(x)$ may be calculated from the following recursive formula

$$G_i(x) = \int_0^x e^{-t^2/2i(i+1)} G_{i-1}(t)\, dt$$

$$G_0(x) = 1$$

However, actual accuracy requirements call for a longer averaging time, i.e. for a higher number of terms. The sum distribution of Eq. (2-126) may then be approximated, according to Eq. (2-69), by a normal distribution having an expected value of

$$M(\eta) = kM(|\xi|) = k\sqrt{\frac{2}{\pi}} \qquad (2\text{-}128)$$

and a variance of

$$D^2(\eta) = kD^2(|\xi|) = k\left(1 - \frac{2}{\pi}\right)\sigma^2 \qquad (2\text{-}129)$$

The relative standard deviation is

$$\frac{D(\eta)}{M(\eta)} = \sqrt{\frac{\pi}{2} - 1}\, \frac{1}{\sqrt{k}} \cong \frac{0.75}{\sqrt{k}} \qquad (2\text{-}130)$$

Finally, for detecting a narrow-band noise with low relative bandwidth, the distribution of the envelope curve should be known. According to Fig. 2-10b, let us select a narrow band of width $\Delta\omega$, which is small compared with ω_0, from the white-noise spectrum shown in Fig. 2-10a. A possible time function of the signal thus generated is shown in Fig. 2-10d: indeed, the signal with centre frequency ω_0 and bandwidth $\Delta\omega$ may be considered as a carrier of frequency ω_0, stochastically modulated in amplitude and phase, and having an instantaneous value of

$$\xi(t) = \varrho(t) \cos\left[\omega_0 t + \vartheta(t)\right] \qquad (2\text{-}131)$$

where $\xi(t)$, $\varrho(t)$ and $\vartheta(t)$ are stochastic time functions; $\varrho(t)$ and $\vartheta(t)$ are the instantaneous values of the envelope and the phase angle, respectively, $(0 < \varrho < \infty$, $0 < \vartheta < 2\pi)$. $\xi(t)$ may also be written in the form

$$\xi(t) = \alpha(t) \cos \omega_0 t + \beta(t) \sin \omega_0 t \qquad (2\text{-}132)$$

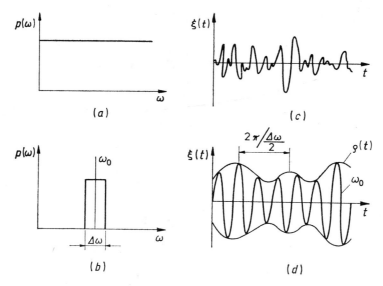

Figure 2-10 White-noise characteristics. *(a)* Power spectrum without band limiting. *(b)* Power spectrum with band limiting. *(c)* Time function without band limiting. *(d)* Time function and envelope function with band limiting.

where $\alpha(t)$ and $\beta(t)$ are mutually independent random variables with normal distributions if the distribution of the instantaneous values pertaining to the wide-band noise of Fig. 2-10c is also normal.

Let us define the two-variable density function $g(R, \theta)$: $g(R, \theta)dR\,d\theta$ is the probability of the instantaneous value of $\xi(t)$ falling within the intervals R and $R+dr$, and θ and $\theta+d\theta$. Let the common density function $f(a, b)$ of $\alpha(t)$ and $\beta(t)$ be similarly defined. The relationship between $g(R, \theta)$ and $f(a, b)$ is calculated with the aid of Fig. 2-11. Thus

$$f(a, b)\Delta a\Delta b = g(R, \Theta)\Delta R\,\Delta\Theta \tag{2-133}$$

As the random variables are independent, $f(a, b)=f(a)f(b)$, so

$$f(a)f(b) = \left(\frac{1}{\sigma\sqrt{2\pi}}\right)^2 e^{-(a^2+b^2)/2\sigma^2} = \frac{1}{2\pi\sigma^2} e^{-R^2/2\sigma^2} \tag{2-134}$$

For the coordinate transformation $a, b \rightarrow R, \Theta$, we have the following relation between the infinitely small surface elements:

$$dRd\Theta = \begin{bmatrix} \dfrac{dR}{da} & \dfrac{dR}{db} \\[2mm] \dfrac{d\Theta}{da} & \dfrac{d\Theta}{db} \end{bmatrix} da\,db = |J|\,da\,db \tag{2-135}$$

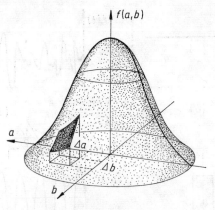

Figure 2-11 Two-variable density function for calculating Rayleigh distribution.[2] (*Courtesy of John Wiley and Sons, Inc.*)

where $|J|$ is the absolute value of the Jacobi determinant. In this case,

$$R = \sqrt{a^2 + b^2}$$

$$\Theta = \arctan \frac{b}{a} \qquad (2\text{-}136)$$

$$|J| = \frac{1}{R}$$

so

$$g(R, \Theta) = \frac{f(a, b)}{|J|} = \frac{R}{2\pi\sigma^2} e^{-R^2/2\sigma^2} \qquad (2\text{-}137)$$

As the distribution of Θ between 0 and 2π is uniform and independent of R, Θ is easily eliminated. Thus

$$f(R) = \int_0^{2\pi} g(R, \Theta)\, d\Theta = 2\pi g(R, \Theta) \qquad (2\text{-}138)$$

From this equation, the single variable density function of the envelope curve **Rayleigh distribution** is

$$\boxed{f(R) = \frac{R}{\sigma^2} e^{-R^2/2\sigma^2} \qquad R \geqq 0} \qquad (2\text{-}139)$$

Integrating the distribution function of Eq. (2-139) between 0 and R, we have

$$F(R) = \mathscr{P}(\varrho < R) = 1 - e^{-R^2/2\sigma^2} \qquad (2\text{-}140)$$

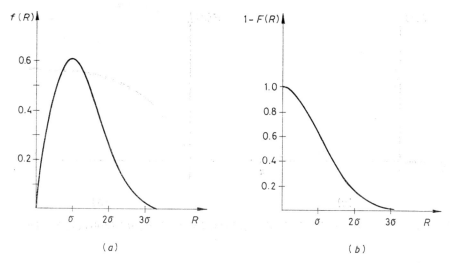

Figure 2-12 *(a)* Density function of Rayleigh distribution. *(b)* Probability of $R>k$.

Thus the probability of an envelope amplitude higher than R is given by $e^{-R^2/2\sigma^2}$ (see Fig. 2-12b). The expected value of the distribution is

$$M(\varrho) = \int_0^\infty \frac{R^2}{\sigma^2} e^{-R^2/2\sigma^2}\, dR = \sqrt{\frac{\pi}{2}}\,\sigma \cong 1.25\sigma \qquad (2\text{-}141)$$

and the variance is

$$D^2(\varrho) = \mu_2 - \mu_1^2 = \int_0^\infty \frac{R^3}{\sigma^2} e^{-R^2/2\sigma^2}\, dR - \mu_1^2 = 2\sigma^2 - \frac{\pi}{2}\sigma^2 = 0.43\sigma^2 \quad (2\text{-}142)$$

Again taking into account the averaging process following the envelope detector, the relative standard deviation of the sum variable formed by the independent, random variables $\varrho_1, \varrho_2, \ldots, \varrho_k$, as in Eqs. (2-123) and (2-130), is

$$\frac{D(\sum_k \varrho)}{M(\sum_k \varrho)} = \frac{\sqrt{0.43}}{1.25}\frac{1}{\sqrt{k}} = \frac{0.525}{\sqrt{k}} \qquad (2\text{-}143)$$

The Rayleigh distribution of Eq. (2-139) may also be derived in a different way.[13]
Another continuous distribution of importance is the exponential distribution that is utilized to describe decomposition, collision, and relaxation processes in atoms and solid state structures.

A continuous random variable ξ has an exponential distribution if the density function has the form

$$f(x) = \begin{cases} 0 & \text{if} \quad x \leq 0 \\ \lambda e^{-\lambda x} & \text{if} \quad x > 0 \end{cases} \qquad (2\text{-}144)$$

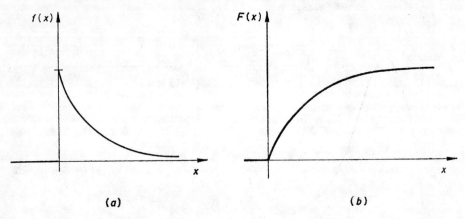

Figure 2-13 Exponential distribution. *(a)* Density function. *(b)* Distribution function.

where λ may be an arbitrary positive number called distribution parameter. The distribution function of ξ, shown in Fig. 2-13*b*, is given by

$$F(x) = \begin{cases} 0 & \text{if} \quad x \leq 0 \\ 1 - e^{-\lambda x} & \text{if} \quad x > 0 \end{cases} \tag{2-145}$$

The first- and second-order moments and the variance of a random variable of exponential distribution and of distribution parameter λ are

$$M(\xi) = \int_0^\infty x \lambda e^{-\lambda x} \, dx = \frac{1}{\lambda} \tag{2-146}$$

$$M(\xi^2) = \int_0^\infty x^2 \lambda e^{-\lambda x} \, dx = \frac{2}{\lambda^2} \tag{2-147}$$

$$D^2(\xi) = M(\xi^2) - M^2(\xi) = \frac{1}{\lambda^2} \tag{2-148}$$

Let us assume that the random variable ξ stands for a time interval and has an exponential distribution. If any time instant x is selected and if the random time interval is not finished at this time instant, the whole process may be regarded as though it had started at the instant x. This property makes the exponential distribution suitable for describing the statistic behavior of elementary particles.

PROBLEMS

2-1 Find the amplitude density functions of the 3 V peak triangle signal and 2 V peak sweep signal shown in Fig. P2-1 and the density function pertaining to the sum of these two signals.

Answer: According to Eq. (1-14),

$$f_1(U_1)\,\Delta U_1 = \frac{\Delta t_1}{T_1} + \frac{\Delta t_2}{T_1}$$

thus

$$f_1(U_1) = \frac{1}{T_1}\left(\frac{dt_1}{dU_1} + \frac{dt_2}{dU_1}\right) = \frac{2}{T_1}\frac{T_1}{4U_{max}} = \frac{1}{2U_{max}} = \frac{1}{6}\ V^{-1}$$

Similarly for the sweep signal

$$f_2(U_2) = \frac{1}{T_2}\frac{dt}{dU_2} = \frac{1}{T_2}\frac{T_2}{2U_{max}} = \frac{1}{2U_{max}} = \frac{1}{4}\ V^{-1}$$

The convolution of the two density functions yields the density function of $U = U_1 + U_2$:

$$f(U) = \begin{cases} \int\limits_{-\infty}^{\infty} f_1(U_1)f_2(U-U_1)\,dU_1 = 0 & U < -5\,V \qquad U > +5\,V \\[2mm] 0.042\left(\dfrac{1}{V^2}\right)(U+5\,V) & -5\,V < U < -1\,V \\[2mm] 0.167\left(\dfrac{1}{V}\right) & -1\,V < U < 1\,V \\[2mm] -0.042\left(\dfrac{1}{V^2}\right)(U-5\,V) & 1\,V < U < 5\,V \end{cases}$$

2-2 Random numbers of normal distribution are required for a computer model of a noise problem. However, only random numbers of uniform distribution, falling in the interval $-1 < \xi < +1$ are available (see Fig. P2-2a). What method may be applied to approximate the normal distribution?

Answer: According to the central limit theorem, the sum of an infinite number of terms has a normal distribution if $\sigma_1 = \sigma_2 = ...\sigma_n$. Let us investigate this for a finite number of terms, and let the number of terms be, in this case, 6! The basic distribution being symmetrical, the sum distribution is also symmetrical, with an expected value of $M(\eta) = 0$ and a variance of

$$D^2(\eta) = 6D^2(\xi) = 6\times\frac{1}{3} = 2 = \sigma^2$$

This is explained by Eq. (2-111), according to which the variance of a uniform distribution between ± 1 is $1/3$.

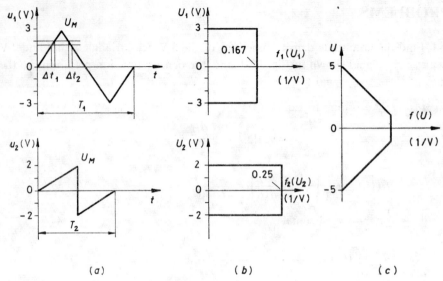

(a) (b) (c)

Figure P2-1 Noncoherent triangular signal and sweep signal *(a)* Time functions. *(b)* Amplitude density functions. *(c)* Density function corresponding to the sum of the two signals.

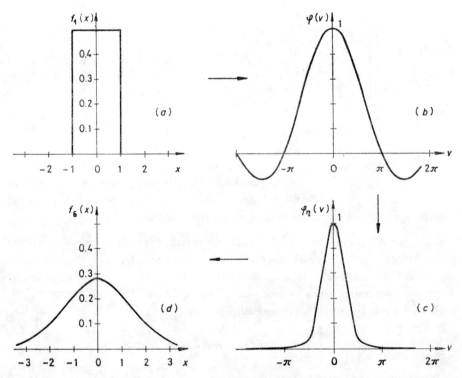

Figure P2-2 *(a)* Even distribution. *(b)* Characteristic function of even distribution. *(c)* Characteristic function on the sixth power. *(d)* Sum distribution.

The density function $f(x)=1/2$ $(-1<\xi<1)$ is symmetrical, so its characteristic function is given, according to Eq. (2-39) (see Fig. P2-2b), by

$$\varphi(v) = \int_{-1}^{1} \cos vx \frac{1}{2} dx = \frac{1}{2} \frac{\sin vx}{v}\bigg|_{-1}^{1} = \frac{\sin v}{v}$$

The characteristic function of an n-term sum, according to Eq. (2-45), is (see Fig. P2-2c)

$$\varphi_n(v) = \varphi^n(v) = \left(\frac{\sin v}{v}\right)^n$$

The density function of the sum is determined by the inverse Fourier transform (see Fig. P2-2d and Appendix B)

$$f(x) = \frac{1}{2\pi} \int_{-\infty}^{\infty} \cos vx \left(\frac{\sin v}{v}\right)^n dv = \frac{1}{\pi} \int_{0}^{\infty} \cos vx \left(\frac{\sin v}{v}\right)^n dv$$

$$= \frac{n}{2^n} \sum_{k=0}^{(x+n)/2} \frac{(-1)^k(n+x-2k)^{n-1}}{k!(n-k)!}$$

For $n=6$ and $-6\leq x<-4$, only the polynomial term pertaining to $k=0$ is left. Thus

$$f(x) = \frac{n}{2^n} \frac{1}{n!}(n+x)^{n-1} \qquad -6 \leq x < -4$$

Further sections of the function are

$$f(x) = \begin{cases} \dfrac{n}{2^n}\left[\dfrac{1}{n!}(n+x)^{n-1} - \dfrac{1}{(n-1)!}(n-2+x)^{n-1}\right] & \text{if} \quad -4 < x < -2 \\[3mm] \dfrac{n}{2^n}\left[\dfrac{1}{n!}(n+x)^{n-1} - \dfrac{1}{(n-1)!}(n-2+x)^{n-1} + \dfrac{1}{2(n-2)!}(n-4+x)^{n-1}\right] \\[2mm] \hspace{6cm} \text{if} \quad -2 \leq x < 0 \end{cases}$$

The polynomial need not be expressed for $x>0$ as $f(x)=f(-x)$. In the following table, the density function thus determined is compared with the density function of a normal distribution with a standard deviation of $\sigma=\sqrt{2}$. For ease of evaluation, both functions are multiplied by the reciprocal value of $1/\sigma\sqrt{2\pi}=1/\sqrt{2}\sqrt{2\pi}$.

Table 2-2

$\pm x$	$2\sqrt{\pi}f_6(x)$	$e^{-x^2/4}$
0	0.9748	1
1	0.7763	0.7788
2	0.3840	0.3679
3	0.1094	0.1054
4	0.0148	0.0183
5	0.00046	0.00193
6	0	0.000123

According to the required accuracy, the number of terms may be increased or decreased. A sum distribution with unity standard deviation may be realized using the definition

$$\eta = \sqrt{\frac{3}{n} \sum_{i=1}^{n} \xi_i}$$

2-3 A p–n junction carries a DC current of 1 μA, and the shot noise is mesaured in a bandwidth of 10 MHz. May the difference between the instantaneous value and the mean value of the current be regarded as having a normal distribution?

Answer: According to Eq. (1-11), which has been derived from the Poisson distribution, the square root taken from the mean square value of the fluctuation, i.e. the standard deviation, is

$$\sqrt{\overline{i^2}} = \sqrt{2qI\Delta f} = \sqrt{3.2 \times 10^{-19} \times 10^{-6} \times 10^7} = 1.8 \times 10^{-9}\ \text{A}$$

As $\sqrt{\overline{i^2}} \ll I$, the Poisson distribution may by substituted by a normal distribution according to Eqs. (2-90) to (2-96). This may not be justified for devices operating at extremely low current and high bandwidth, e.g., for optoelectronic devices.

2-4 The narrow-band noise of normal distribution, shown in Fig. 2-10d is used to feed two detectors—an absolute-value detector and an envelope detector—and the detected signals are integrated by a high time-constant RC network (Fig. P2-3). Will the two output voltages be different?

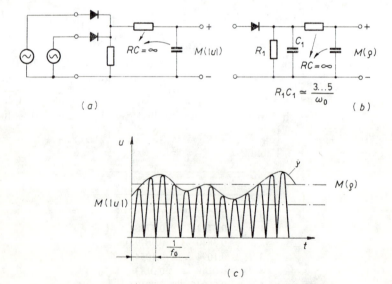

(a)

(b)

(c)

Figure P2-3 *(a)* Absolute-value detector. *(b)* Envelope detector. *(c)* Absolute value of narrow-band noise voltage.

Answer: Let the root mean square value of the input noise voltage be

$$U_{\text{rms}} = \sqrt{\overline{u^2}} = D(u) = \sigma$$

Then, according to Eq. (2-69), the DC output voltage of the absolute-value detector will be

$$M(|u|) = \sqrt{\frac{2}{\pi}}\, U_{\text{rms}}$$

On the other hand, the DC output voltage of the envelope detector, according to Eq. (2-141), will correspond to the expected value of the Rayleigh distribution

$$M(\varrho) = \sqrt{\frac{\pi}{2}}\, U_{\text{rms}} = \frac{\pi}{2}\, M(|u|)$$

or

$$M(|u|) = \frac{2}{\pi}\, M(\varrho)$$

Taking into account the approximate sinusoidal waveform of the narrow-band noise, the output of the absolute-value detector will be a series of half sine waves, with amplitudes corresponding to the envelope function. The mean value of this half sine-wave series is $2/\pi$ times the peak value.

2-5 Show that the density function χ^2 of the distribution (2-118) meets the requirement

$$\int_{-\infty}^{\infty} f(y)\, dy = 1$$

following from the definition of the density function, e.g. for $k=4$.

    Answer: For integers, $\Gamma(\alpha+1)=\alpha!$, $\Gamma(2)=1$, moreover $f_k(y)=0$ for $y<0$, so with partial integration, we have

$$\int_0^\infty f_4(y)\, dy = \frac{1}{4\sigma^4} \int_0^\infty e^{-y/2\sigma^2}\, dy = \frac{1}{4\sigma^4}\left[-2\sigma^2 y e^{-y/2\sigma^2} - 4\sigma^4 e^{-y/2\sigma^2}\right]_0^\infty = 1.$$

REFERENCES

[1] Arend, K.: "Some Combinations of Noise Signals", *Proc. IEEE*, vol. 50, no. 12, pp. 2518–2519, 1962.

[2] Bendat, J. S. and A. G. Piersol: *Measurement and Analysis of Random Data*, John Wiley, New York, 1966.

[3] Farkas, M.: *Special Functions* (Speciális függvények), Műszaki Könyvkiadó, Budapest, 1964, (in Hungarian).

[4] Feller, W.: *An Introduction to Probability Theory and Its Applications*, John Wiley, New York, vol. 1, 1957; vol 2, 1966.

[5] Gnedenko, B. V.: *The Theory of Probability*, Mir, Moscow, 1969; 1973; 1976.

[6] Gnedenko, B. V. and A. N. Kolmogoroff: *Limit Distributions for Sums of Independent Variables*, Addison-Wesley, Reading, 1954.

[7] Godwin, H. J.: "On the Distribution of the Estimate of Mean Deviation", *Biometrika*, vol. 33, no. 3, pp. 254—256, 1945.

[8] Hubbard, W. M.: "The Approximation of a Poisson Distribution by a Gaussian Distribution", *Proc. IEEE*, vol. 58, no. 9, pp. 1374—1375, 1970.

[9] Jahnke, E. and F. Emde: *Tables of Functions with Formulae and Curves*, Dover, New York, 1945.

[10] Miller, K. S.: *Engineering Mathematics*, Dover, New York, 1963.

[11] Papoulis, A.: *Probability, Random Variables and Stochastic Processes*, McGraw-Hill, New York, 1965.

[12] Pfeiffer, P. E.: *Concepts of Probability Theory*, McGraw-Hill, New York, 1965.

[13] Powell, A.: "On the Fatigue Failure of Structures Due to Vibrations Excited by Random Pressure Fields", *JASA*, vol. 30, no. 12, pp. 1130—1135, 1958.

[14] Prékopa, A.: *Theory of Probability* (Valószínűségelmélet), Műszaki Könyvkiadó, Budapest, 1962, (in Hungarian).

[15] Rényi, A.: *Probability Calculus* (Valószínűségszámítás), Tankönyvkiadó, Budapest 1968, (in Hungarian).

[16] Schwarze, H.: "Eine Untersuchung zum zentralen Grenzwertsatz der Wahrscheinlichkeitstheorie", *NTZ*, vol. 23, no. 12, pp. 640—642, 1970.

[17] *Electronics*, vol. 50, no. 20, p. 114, 1977.

[18] "The Fast Fourier Transform" *IEEE Trans.*, vol. AU-17, no. 2, 1969.

THREE

STOCHASTIC SIGNALS IN THE TIME AND FREQUENCY DOMAIN

In Chapter 2, the stochastic process has been defined as a single parameter assembly of random variables with the variable covering a time instant set. Thus $\xi=\xi(t)$ may be regarded as the discrete or continuous function of time. In our case, the specific value of ξ may be voltage or current.

Normally, the time function of voltage or current may be expressed by a simple and unique mathematical formula. Time functions of precise time dependence are called deterministic functions. In the case of these functions, a few parameters such as amplitude, frequency and phase are sufficient to describe either the past or the future of the signal. Evidently, the "past" or the "future" is restricted to the time interval of the signal; this time interval has to be defined for each specific case. On the other hand, the past knowledge of a stochastic signal does not allow the precise determination of the next instantaneous value; instead its probability may be ascertained. In other words, an ideal stochastic signal has no "memory". However, there is always some kind of time constant in physically realizable devices and circuits (e.g., the transit time of the charge carrier in a vacuum or solid state device). Owing to this effect, the consecutive instantaneous values of the noise appearing at the output of a device or circuit, assuming infinitely small time intervals between these instantaneous values, are not completely independent from each other. This internal property—autocorrelation—of real stochastic signals will be frequently utilized in this Chapter. However, the time and frequency domain description of deterministic and stochastic signals will first be presented systematically.

3-1 DETERMINISTIC AND STOCHASTIC SIGNALS[1, 2, 3, 4, 7, 9, 10, 11]

Figure 3-1 shows all possible types of signals. Both main groups may be divided further into classes and subclasses. Wherever possible, the usual description in the time, frequency, and complex frequency domains will be presented, as well as the transformations between these domains.

The simplest deterministic signal is the periodic signal with a time function comprising sine and cosine terms, but only with integral multiples of ω_0 in the arguments of this function. The discrete version of the Fourier transform (the Fourier series) is applied for the passing over from the time domain to the frequency domain.

The "almost periodical" signal is an extension of the periodical signal comprising harmonic components. It may be regarded as a superposition of discrete and fixed-frequency sine and cosine terms. However, the quotients of these discrete frequencies are neither integers nor rational fractions, i.e.

$$\frac{\omega_j}{\omega_k} = \text{irrational number}$$

meaning that the signal waveform is repeated only after an infinite time. Such a signal is, for instance, the superposition of the triangular wave and the sweep signal in Problem 2-1, provided that their fundamental frequencies are not harmonically related, according to the above condition. Separately, both signals may be represented in the frequency domain by a discrete spectrum. The connection between the time domain and the frequency domain is given by the Fourier integral, resulting in a spectrum comprising a series of Dirac-delta functions.

Transients, which may be described by a starting function $x(t)=0$ if $t<0$, may also be classified as deterministic functions. As a transient may not repeat itself, i.e. it has an infinite period, the Fourier transform may be used to change over from the time to the frequency domain. Unfortunately, the Fourier transform in itself is not necessarily convergent; the conditions for convergence would have to be investigated for each case. The Laplace transform is much more suitable for the investigation of deterministic transient signals. In spite of this, the Fourier transform will be applied in this book as, with due caution, it may be equally useful for nonperiodic, deterministic signals and for stochastic signals.

Stochastic signals are comparable to the almost periodical signals, with the difference that the frequency spectrum of stochastic signals is, at least piecewise, continuous. The main problem is that neither the amplitude nor the phase can be predicted for an arbitrary time instant, so the amplitude and phase of the Fourier transform are unknown. Stochastic signals should therefore be described by other methods.

A stochastic process may be defined by a function $\xi(t)$ in the same way as a deterministic process. The function values may be real or complex numbers, or

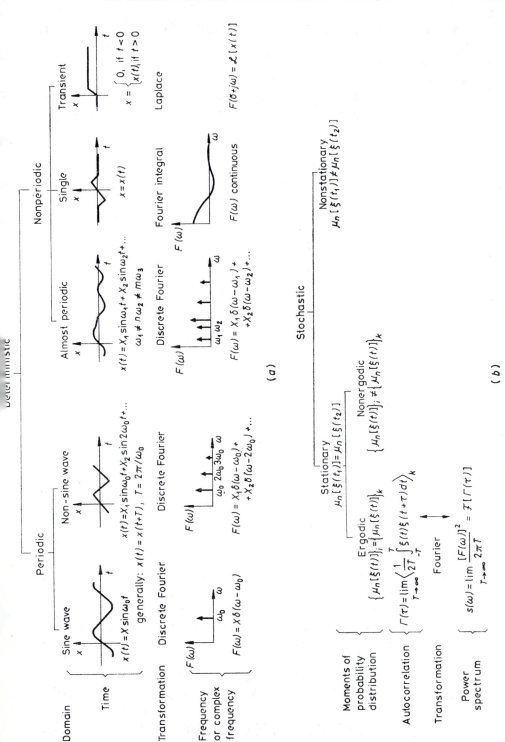

Figure 3-1 Classification of (a) deterministic and (b) stochastic signals.

vectors; t is the argument of the function covering the element values of a set T. A specific function $\xi(t)$ observed during a particular trial is called a realization of the stochastic process.

With a fixed argument t, $\xi(t)$ may have different values during consecutive trials. In other words, $\xi(t)$ may have random values for a fixed t. In such a case, it may naturally be assumed that $\xi(t)$ is a random variable pertaining to a fixed t, in the sense of the probability theory. These random variables form a set—a so-called ensemble.

One way of describing a stochastic signal is by the presentation of the distribution function which has been investigated in detail in Chapter 2. Two stochastic functions, assuming that other conditions are met, may only be regarded as equivalent functions if the consecutive density functions of their time instant values are equal. This would complicate the mathematical treatment, so the density functions are characterized by their moments μ_n.

The density functions may be defined in two ways: either a single realization is considered and the time is varied, or a set of realizations is considered at a fixed time instant. If the first-order moments are then calculated, these may be called time or set averages. A similar distinction may be applied for higher-order moments. For a case in which density functions or moments calculated in two different ways are equal, we speak of an *ergodic process*.

Now we may define the difference between stationary and nonstationary stochastic signals. For a strict-sense stationary signal, all moments of the distribution are independent of time. For a wide-sense stationary signal, only the first- and second-order moments are independent of time. If not even this condition is met, the process is nonstationary. Because of its complexity, this case will not be dealt with.

An *ergodic group* is made up of distributions having equal, same-order moments. On the other hand, if several distributions have moments that are all time-independent but different then all distributions are strict-sense stationary in themselves but do not form an ergodic group.

In Fig. 3-1, the condition for ergodicity is illustrated by comparing moments of arbitrary orders. With a few exceptions, stochastic processes in electronic devices are stationary and ergodic, so these attributes will not be used below.

The first-order moment of the distribution gives the expected value, and the central second-order moment gives the mean square value of the fluctuation, called variance. The latter has a dimension of V^2 or A^2 and thus is directly connected to the *power* of the stochastic signal. On the other hand, real power may be converted into heat and so may be defined for an arbitrary waveform involving only the loss of the frequency and phase information. Thus for a stochastic signal, instead of using the first right-hand term of the expression

$$D^2[\xi(t)] = M[\xi^2(t)] - M^2[\xi(t)] \tag{3-1}$$

which really means the long-term average value of $\xi(t) \times \xi(t)$, the following so-called autocorrelation function is used:

$$\Gamma_\xi(\tau) = \lim_{T \to \infty} \frac{1}{2T} \int_{-T}^{T} \xi(t)\xi(t+\tau)\,dt \qquad (3\text{-}2)$$

Evidently, this is equal to the expected value of $\xi^2(t)$ only for $\tau=0$. However, it has the advantage that by varying the time shift τ, Eq. (3-2) shows the connection—i.e. the autocorrelation—between consecutive time instants. The definition given in Fig. 3-1 may also be applied to nonergodic processes; in these cases, the time mean values according to Eq. (3-2) may be separately calculated for all elementary processes, and then the mean value of the whole assembly taken.

Generally, the stochastic signal occupies a wide frequency range. Although the Fourier transform of the time function is not known, a frequency domain description exists; this is the power spectrum mentioned in Chapter 1. It follows from the theorem of energy conservation that the total power falling within the frequency range $0<f<\infty$ has to be equal to the power calculated from the mean square value of time instants that has to be calculated in a form averaged for the time interval $T \gg 1/f_{min}$. It is thus clear that the power spectrum and some characteristic of the time instants are related. (We shall see in Sec. 3-3 that a mutual connection between the power spectrum and the autocorrelation function is given by the Fourier transform.) Naturally, the previous definitions may also be applied for deterministic signals; this is not shown in Fig. 3-1. In Sec. 3-2 the properties of the power spectrum will be investigated in detail.

3-2 POWER SPECTRUM

For simplicity, we shall begin with a deterministic signal, or a periodic non-sinusoidal waveform. The time function of this waveform may be expressed by a Fourier series. Thus

$$x(t) = \frac{a_0}{2} + \sum_{n=1}^{\infty}(a_n \cos n\omega_0 t + b_n \sin n\omega_0 t) \qquad (3\text{-}3)$$

where

$$a_0 = \frac{2}{T} \int_{-T/2}^{T/2} x(t)\,dt \qquad (3\text{-}4)$$

$$a_n = \frac{2}{T} \int_{-T/2}^{T/2} x(t) \cos n\omega_0 t\,dt \qquad (3\text{-}5)$$

$$b_n = \frac{2}{T} \int_{-T/2}^{T/2} x(t) \sin n\omega_0 t\,dt \qquad (3\text{-}6)$$

and

$$T = \frac{2\pi}{\omega_0} \tag{3-7}$$

Utilizing the Euler relation, we have

$$\cos n\omega_0 t = \frac{e^{jn\omega_0 t} + e^{-jn\omega_0 t}}{2} \tag{3-8}$$

$$\sin n\omega_0 t = \frac{e^{jn\omega_0 t} - e^{-jn\omega_0 t}}{2j} \tag{3-9}$$

and Eq. (3-3) transforms to

$$x(t) = \frac{a_0}{2} + \sum_{n=1}^{\infty} \left(\frac{a_n + jb_n}{2} e^{-jn\omega_0 t} + \frac{a_n - jb_n}{2} e^{jn\omega_0 t} \right) \tag{3-10}$$

Let us now introduce the complex coefficients

$$c_n^- = \frac{a_n + jb_n}{2} \tag{3-11}$$

and

$$c_n^+ = \frac{a_n - jb_n}{2} \tag{3-12}$$

Thus $x(t)$ is made up of a constant term plus the sum of vectors rotating with angular velocities $n\omega_0$ and $-n\omega_0$. The absolute value and phase of these vectors are given by coefficients c_n^- and c_n^+. In the following, let $a_0 = 0$. The introduction of the seemingly unjustified negative angular frequency $-n\omega_0$ results in significant mathematical simplifications. Note that

$$c_n^- = (c_n^+)^* \tag{3-13}$$

i.e. the corresponding coefficient pair members are conjugates of each other. Let us substitute $-n\omega_0$ instead of $n\omega_0$ into the defining formulas (3-5) and (3-6) for a_n and b_n, respectively; then b_n changes sign while a_n does not change. Thus in the range $-\infty < n < \infty$, we may apply the universal coefficient

$$c_n = \frac{a_n - jb_n}{2} = \frac{1}{T} \int_{-T/2}^{T/2} x(t) e^{-jn\omega_0 t} \, dt \tag{3-14}$$

instead of the separate coefficients c_n^- and c_n^+. Thus $x(t)$ may be written in the simplified form

$$x(t) = \sum_{-\infty}^{\infty} c_n e^{jn\omega_0 t} \tag{3-15}$$

Expressing c_n as given in Eq. (3-14), Eq. (3-15) will comprise a vector pertaining to increasing time.

Let us now calculate the mean square value of $x(t)$. To this end, the multi-term right-hand side expression in Eq. (3-15) has to be multiplied by its own conjugated value and then has to be averaged in time. Thus

$$M[x^2(t)] = \frac{1}{T} \int_{-T/2}^{T/2} (c_1 c_1^* + c_1 e^{j\omega_0 t} c_2^* e^{-j2\omega_0 t} + \ldots + c_2 e^{j2\omega_0 t} c_1^* e^{-j\omega_0 t} + c_2 c_2^* + \ldots) \, dt$$

$$= \frac{1}{T} \int_{-T/2}^{T/2} \sum_n \sum_m c_n c_m^* e^{j(n-m)\omega_0 t} \, dt \qquad (3\text{-}16)$$

The period $T = 2\pi/\omega_0$ of the lowest frequency component is chosen as average time. In the exponent of the exponential term, we find n and m, and t as factors of a single product, so the sequence of operations involving these factors may be inverted. Thus

$$M[x^2(t)] = \frac{1}{T} \sum_n \sum_m \left(c_n c_m^* \int_{-T/2}^{T/2} e^{j(n-m)\omega_0 t} \, dt \right) \qquad (3\text{-}17)$$

For $n \neq m$, we have

$$\int_{-T/2}^{T/2} e^{j(n-m)(2\pi t/T)} \, dt = 0$$

so finally

$$M[x^2(t)] = \sum_{n=-\infty}^{\infty} c_n c_n^* = \sum_{n=-\infty}^{\infty} |c_n|^2 \qquad (3\text{-}18)$$

However, according to Eq. (3-13), the coefficients c_n^+ and c_n^- pertaining to the same number n are conjugates of each other, thus their absolute values are equal. It follows therefore that $|c_n|^2$ is an even function, which provides substantial computational benefits.

Figure 3-2a shows the complex coefficients c_n and the spectrum distribution of the mean square value pertaining to an arbitrarily chosen deterministic function. Figure 3-2b cannot be regarded truly as a power spectrum as the dimension of coefficients c_n is V or A, thus the dimension of $|c_n|^2$ is V^2 or A^2. Actually, we may call this a power spectrum only in cases where this voltage appears across, or this current flows through, a given resistance R. Unless this leads to ambiguity, the coefficients $|c_n|^2$ will be called the terms of the power spectrum. However, it should be noted that in calculating the overall power of the signal $x(t) = i(t)$, Eq. (3-18) has to be multiplied by the value of resistance R. Thus

$$P = R \sum_{n=-\infty}^{\infty} |c_n|^2 \; \text{(W)} \qquad (3\text{-}19)$$

Equation (3-19) expresses a simple theorem: the power of a periodic nonsinusoidal wave may be calculated by taking into account all Fourier coefficients.

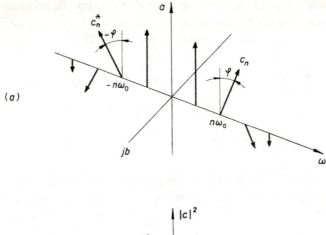

(a)

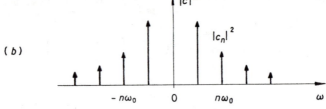

(b)

Figure 3-2 Periodic nonsinusoidal signal. *(a)* Complex Fourier coefficients as a function of discrete $n\omega_0$ frequencies. *(b)* Double-sided power spectrum.

The use of negative frequencies is not convenient from the measurement standpoint. As $|c_n|^2$ is an even function and according to our condition $c_0=a_0=0$, the following relation also holds:

$$M[x^2(t)] = 2 \sum_{n=1}^{\infty} |c_n|^2 = \sum_{n=1}^{\infty} |C_n|^2 \qquad (3\text{-}20)$$

where $|C_n|^2=2|c_n|^2$. In the following, lower case letters will be used to denote members of the two-sided power spectrum, while upper case letters denote members of the single-sided power spectrum which is only interpreted for positive frequencies.

Let us now investigate a deterministic signal with a continuous Fourier spectrum (see Fig. 3-4a). For this case, the discrete spectrum made up of coefficients c_n is changed into a continuous spectrum $F(\omega)$, using the following relation which is analogous to Eq. (3-14):[5, 6, 8]

$$F(\omega) = \int_{-\infty}^{\infty} x(t)e^{-j\omega t}\,dt \qquad (3\text{-}21)$$

As Eq. (3-21) is an improper integral and $|e^{-j\omega t}|=1$, the following conditions have to be met for a finite value of $x(t)$:

1. $x(t)$ should be a continuous function capable of differentiation.
2. The integral of the absolute value of $x(t)$ should be finite, i.e.

$$\int_{-\infty}^{\infty} |x(t)|\, dt = \text{finite} \tag{3-22}$$

This means that for $t \to \pm\infty$, $x(t) \to 0$. As this last condition would substantially reduce the applicable function types—and, as will be shown, the stationary stochastic processes would also be excluded—let, for the time being, the finite value of Eq. (3-21) be secured by the condition that the function $x(t)$ may differ from zero only in the interval between $-T$ and $+T$ (see Fig. 3-3). Thus

$$x(t) = \begin{cases} x_T(t) & \text{if} \quad -T < t < T \\ 0 & \text{otherwise} \end{cases} \tag{3-23}$$

Equation (3-21) may then be written as

$$F_T(\omega) = \int_{-\infty}^{\infty} x_T(t) e^{-j\omega t}\, dt = \int_{-T}^{T} x(t) e^{-j\omega t}\, dt \tag{3-24}$$

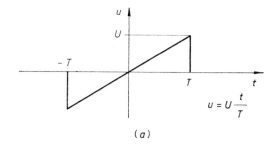

(a)

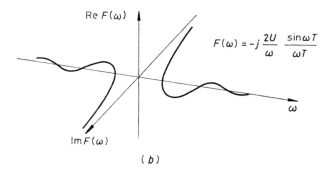

$$F(\omega) = -j\,\frac{2U}{\omega}\,\frac{\sin \omega T}{\omega T}$$

(b)

Figure 3-3 Nonperiodic signal truncated in time.
(a) Time function. (b) Fourier spectrum.

Due to the requirement that $x(t)$ be real, Eq. (3-24) may be divided into two parts:

$$F_T(\omega) = \int_{-\infty}^{\infty} x_T(t) \cos \omega t \, dt - j \int_{-\infty}^{\infty} x_T(t) \sin \omega t \, dt \qquad (3\text{-}25)$$

Writing $(-\omega)$ instead of ω, the first right-hand side term is unchanged while the second term changes sign. Thus

$$F_T(-\omega) = F_T^*(\omega) \qquad (3\text{-}26)$$

Knowledge of $F_T(\omega)$ allows the determination of $x_T(t)$ by inverse Fourier transform

$$x_T(t) = \frac{1}{2\pi} \int_{-\infty}^{\infty} F_T(\omega) e^{j\omega t} \, d\omega \qquad (3\text{-}27)$$

Let us now calculate the time integral of $x_T^2(t)$, which is proportional to the signal energy. According to condition (3-22) we have to arrive at a finite result

$$\int_{-\infty}^{\infty} x_T^2(t) \, dt = \int_{-\infty}^{\infty} x_T(t) x_T(t) \, dt = \int_{-\infty}^{\infty} x_T(t) \left[\frac{1}{2\pi} \int_{-\infty}^{\infty} F_T(\omega) e^{j\omega t} \, d\omega \right] dt \quad (3\text{-}28)$$

where Eq. (3-27) has also been used. Note that in the double integral of Eq. (3-28), $x_T(t)$ and $F_T(\omega)$ are symmetrically placed, and $e^{j\omega t}$ comprises both integral variables with equal weights. So as in Eqs. (3-16) and (3-17), the sequence of integrations here may also be exchanged. Thus

$$\int_{-\infty}^{\infty} x_T^2(t) \, dt = \frac{1}{2\pi} \int_{-\infty}^{\infty} F_T(\omega) \left[\int_{-\infty}^{\infty} x_T(t) e^{j\omega t} \, dt \right] d\omega$$

$$= \frac{1}{2\pi} \int_{-\infty}^{\infty} F_T(\omega) F_T(-\omega) \, d\omega \qquad (3\text{-}29)$$

because the internal integral is the Fourier transform of $x_T(t)$ as a function of $-\omega$. Utilizing Eq. (3-26), we have

$$\int_{-\infty}^{\infty} x_T^2(t) \, dt = \frac{1}{2\pi} \int_{-\infty}^{\infty} |F_T(\omega)|^2 \, d\omega \qquad (3\text{-}30)$$

This is the Parseval theorem.

Note that without the condition of Eq. (3-23), both sides of Eq. (3-30) would go to infinity for $T \to \infty$ because the energy of a finite amplitude real signal is infinite for an infinite time. It is thus advisable to use the mean square value which is proportional to the time average power; this may be finite even in this case:

$$M[x_T^2(t)] = \frac{1}{2T} \int_{-T}^{T} x_T^2(t) \, dt = \frac{1}{4\pi T} \int_{-\infty}^{\infty} |F_T(\omega)|^2 \, d\omega \qquad (3\text{-}31)$$

Let us remove the restriction imposed by Eq. (3-23) and now we only require that $M[x^2(t)]$ should go to a finite limiting value for $T \to \infty$. Then

$$M[x^2(t)] = \lim_{T \to \infty} \frac{1}{4\pi T} \int_{-\infty}^{\infty} |F(\omega)|^2 \, d\omega \qquad (3\text{-}32)$$

If $x(t)$ has current or voltage dimension, the average power may be defined as

$$P = R^{\pm 1} M[x^2(t)] \qquad (3\text{-}33)$$

where $[R] = V/A$.

The power pertaining to one radian per second which is the *power density* is given by

$$p(\omega) = \frac{dP}{d\omega} = R^{\pm 1} \lim_{T \to \infty} \frac{1}{4\pi T} |F(\omega)|^2 \qquad (3\text{-}34)$$

which now may be accurately defined with the aid of Eq. (3-34). Do not forget, however, that this definition applies to a two-sided power spectrum, i.e.

$$P = \int_{-\infty}^{\infty} p(\omega) \, d\omega \qquad (3\text{-}35)$$

With some degree of inaccuracy, the quantity

$$s(\omega) = \frac{p(\omega)}{R^{\pm 1}} = \lim_{T \to \infty} \frac{1}{4\pi T} |F(\omega)|^2 \qquad -\infty < \omega < \infty \qquad (3\text{-}36)$$

which has a dimension of V^2/rad or A^2/rad, is also called the power density, although denoted by another letter, s. It follows from Eq. (3-26), as in Eq. (3-20), that $|F(\omega)|^2$ is an even function, so the one-sided power density function may again be introduced. Thus

$$S(\omega) = 2s(\omega) = \lim_{T \to \infty} \frac{1}{2\pi T} |F(\omega)|^2 \qquad 0 < \omega < \infty \qquad (3\text{-}37)$$

Instead of $S(\omega)$ and $s(\omega)$, $S(f)$ and $s(f)$ are used in many cases. These terms are evidently related.

$$\int_{-\infty}^{\infty} s(\omega) \, d\omega \equiv \int_{-\infty}^{\infty} s(f) \, df \qquad (3\text{-}38a)$$

and

$$\int_{0}^{\infty} S(\omega) \, d\omega \equiv \int_{0}^{\infty} S(f) \, df \qquad (3\text{-}38b)$$

As $d\omega = 2\pi df$, we have

$$s(f) = 2\pi s(\omega) = \lim_{T \to \infty} \frac{1}{2T} |F(\omega)|^2 \qquad \text{where} \qquad -\infty < f < \infty \qquad (3\text{-}39)$$

and

$$S(f) = 2\pi S(\omega) = \lim_{T \to \infty} \frac{1}{T} |F(\omega)|^2 \qquad \text{where} \qquad 0 < f < \infty \qquad (3\text{-}40)$$

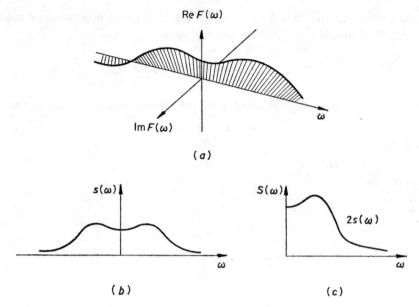

Figure 3-4 Nonperiodic, deterministic signal. *(a)* Continuous Fourier spectrum. *(b)* Double-sided power density function. *(c)* Single-sided power density function

Figure 3-4 shows a continuous Fourier spectrum, as well as a two-sided and a one-sided power density function. However, it is important to note that while $F(\omega)$ is a complex function with a real variable, so for each ω it defines a pair of real–imaginary values; $s(\omega)$ is a real function with one variable. When calculating $s(\omega)$, the phase information is lost, so $F(\omega)$ may not be determined in a unique sense from $s(\omega)$.

The mean square value may also be defined for stochastic processes, so utilizing Eq. (3-33) the average power of a stochastic signal may also be determined. It is also true that this power is not related to a single frequency as all random processes involve several components of different frequencies. Thus the power density spectrum may also be applied for a stochastic signal, i.e.

$$M[\xi^2(t)] = \int_{-\infty}^{\infty} s(\omega)\,d\omega \qquad (3\text{-}41)$$

$s(\omega)$ may also comprise Dirac pulses. However, the chief problem is not the impossibility of computing the Fourier spectrum back from this, but that $F(\omega)$ may not exist at all. [It is not finite, so the condition of Eq. (3-22) is not met. Simultaneously, the Laplace transform may exist.] For instance, consider a stationary stochastic signal of infinite time duration. The requirement of infinite time duration will make Eq. (3-24) an improper integral, and the stationary property will not allow the condition of Eq. (3-22) to be met. This is explained by the fact that when only $\xi(t)\not\equiv 0$ then

$$\int_{-\infty}^{\infty} |\xi(t)|\, dt \to \infty \tag{3-42}$$

However, $s(\omega)$ may be defined even under these circumstances. The stationarity condition will render $M[\xi^2(t)]=\mu_2[\xi(t)]$ a finite, time-independent constant, so the right-hand side of Eq. (3-41) will also be finite. Naturally, this will not exclude the possibility that very high amplitude, but very narrow pulses (Dirac deltas), appear in the spectrum $s(\omega)$.

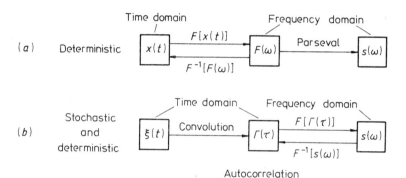

Figure 3-5 Characterizations of *(a)* deterministic, and *(b)* stochastic signals, with permissible transformations.

Figure 3-5a shows the time- and frequency-domain descriptions of the deterministic signals, together with the allowed transforms. The Fourier transform of a stochastic signal may not exist even if the time function $\xi(t)$ is known, and this may prevent the calculation of $s(\omega)$. This necessitates the introduction of a new type of description, the autocorrelation function (see Fig. 3-5b) which may equally be defined both for stochastic and for deterministic signals.

3-3 CORRELATION, AUTOCORRELATION

In our previous calculations, the independence of the random variables has been generally assumed. However, this may not always be the case, so the measure of dependence should somehow be defined. If the two random variables in question are ξ and η, then their relation may be expressed as

$$C_{\xi\eta} = M\{[\xi-M(\xi)][\eta-M(\eta)]\} \tag{3-43}$$

In Eq. (2-34) it has been utilized that for independent random variables $C_{\xi\eta}=0$. Performing the prescribed operations, we have

$$
\begin{aligned}
C_{\xi\eta} &= M(\xi\eta) - 2M(\xi)M(\eta) + M(\xi)M(\eta) \\
&= M(\xi\eta) - M(\xi)M(\eta)
\end{aligned}
\tag{3-44}
$$

If $\eta=\xi$, we once more have the much-used relation [see Eq. (2-26)]

$$
C_{\xi\xi} = D^2(\xi) = M(\xi^2) - M^2(\xi)
$$

Let us now introduce the relation*

$$
\Gamma_{\xi\eta} = M(\xi\eta) = C_{\xi\eta} + M(\xi)M(\eta)
\tag{3-45}
$$

and let us restrict ourselves to calculating $\Gamma_{\xi\eta}$. If ξ and η have nonzero expected values (they have DC components), the product of the DC terms $M(\xi)M(\eta)$ has simply to be subtracted from $\Gamma_{\xi\eta}$. If ξ and η are random variables of two stochastic processes, then both are time functions. Extremely useful information regarding the process structure is gained if the expected value calculation of Eq. (3-45) is performed with different time shifts τ between the two functions. The result thus calculated will be a function of τ. If $\xi(t)$ and $\eta(t)$ are continuous, then

$$
\Gamma_{\xi\eta}(\tau) = M[\xi(t)\eta(t+\tau)] = \lim_{T\to\infty} \frac{1}{2T} \int_{-T}^{T} \xi(t)\eta(t+\tau)\,dt
\tag{3-46}
$$

is the so-called cross-correlation function. However, it is also possible to compare the value at time instant t of a single stochastic process with the value measured earlier by a time interval τ, and to take the average for the whole time scale. We thus arrive to the autocorrelation function

$$
\Gamma_{\xi}(\tau) = M[\xi(t)\xi(t+\tau)] = \lim_{T\to\infty} \frac{1}{2T} \int_{-T}^{T} \xi(t)\xi(t+\tau)\,dt
\tag{3-47}
$$

Naturally it is necessary to choose $T\gg\tau$. It is clear that if there is no correlation between $\xi(t)$ and $\xi(t+\tau)$, not even for infinitely small τ, the sequentially following instantaneous values are totally independent then $\Gamma_{\xi}(\tau)\equiv 0$. On the other hand, the longer lasting the effect of the instantaneous values the slower $\Gamma_{\xi}(\tau)$ tends to zero for $t\to\infty$. The autocorrelation function, being of greater significance, will be treated in more detail.

If the stochastic process is stationary, $\Gamma_{\xi}(\tau)$ is independent of time, so the integration interval between $-T$ and $+T$ may be chosen at any point. Equation (3-47) is valid for stationary ergodic processes. For nonergodic processes, the reasoning of Sec. 3-1 requires the additional averaging of the autocorrelation functions pertaining to partial processes due to various sources.

---

* In the literature, there is no unified terminology for $\Gamma_{\xi\eta}$ and $C_{\xi\eta}$. $\Gamma_{\xi\eta}$ is usually called correlation and $C_{\xi\eta}$ is usually called covariance.

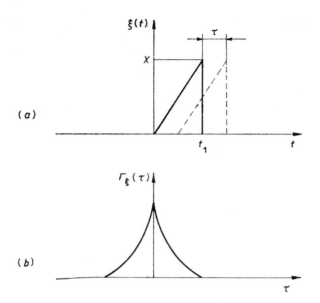

Figure 3-6 Asymmetrical stationary triangular wave.
(a) Time function. (b) Autocorrelation function.

Figure 3-6b shows the characteristic properties of $\Gamma_\xi(\tau)$:

$$\Gamma_\xi(-\tau) = \Gamma_\xi(\tau) \tag{3-48}$$

$$|\Gamma_\xi(\tau)| \leq \Gamma_\xi(0) \tag{3-49}$$

Equation (3-48) can simply be proved. Let us write $\vartheta \to \tau - t$ instead of t in Eq. (3-47), then $d\vartheta = dt$ and

$$\Gamma_\xi(-\tau) = \lim_{T \to \infty} \frac{1}{2T} \int_{-T}^{T} \xi(\vartheta - \tau)\xi(\vartheta)\, d\vartheta = \Gamma_\xi(\tau) \tag{3-50}$$

as for a stationary process, Γ_ξ is dependent only on the time shift τ and not on t or ϑ. Γ_ξ is always an even function independent of the shape of $\xi(t)$. The second relationship, Eq. (3-49), may be proved as follows. Let us calculate the expected value

$$M\{[\xi(t) \pm \xi(t+\tau)]^2\} = M\{[\xi^2(t) \pm 2\xi(t)\xi(t+\tau) + \xi^2(t+\tau)]\} \geq 0 \tag{3-51}$$

This can only be zero or positive, ξ being real. However, for a stationary process $M[\xi^2(t)] = M[\xi^2(t+\tau)]$ so

$$M[\xi^2(t)] \geq M|\xi(t)\xi(t+\tau)|. \tag{3-52}$$

It can be seen that the right-hand side equals $|\Gamma_\xi(\tau)|$ according to Eq. (3-47), and the left side is

$$\Gamma_\xi(0) = M[\xi^2(t)] = \lim_{T \to \infty} \frac{1}{2T} \int_{-T}^{T} \xi^2(t)\, dt = \int_{-\infty}^{\infty} x^2 f(x)\, dx \tag{3-53}$$

This brings us, in addition to the proof of Eq. (3-49), to an important result: the autocorrelation function yields the mean square value of a stochastic or deterministic signal at $\tau=0$. Thus $\Gamma_\xi(\tau)$ includes an earlier result as a particular case and provides more detail about the internal properties of the signal. For complete description in Eq. (3-53), the squared expected value is also expressed by the second-order moment of the distribution [see Eq. (2-30)].

For a signal not comprising periodic components, $\xi(t)$ and $\xi(t+\tau)$ are independent if $\tau\to\infty$. It thus follows that

$$\lim_{\tau\to\infty} \Gamma_\xi(\tau) = 0 \tag{3-54}$$

If increasing τ results in a substantial increase in $\Gamma_\xi(\tau)$, this shows the presence of a periodic component. The interval $\pm\tau_k$ in which $\Gamma_\xi(\tau)$ does not decrease permanently below a prescribed fraction of $\Gamma_\xi(0)$ is called the correlation interval.

This zero convergence property of the autocorrelation function in the case of nonperiodic signals, further the expression according to Eq. (3-47), which is really a convolution integral, makes the Fourier transform feasible. As a starting point let us reconsider a stochastic or deterministic signal, limited in time, again defined by Eq. (3-23). Then, requiring that $\tau\ll T$, we have

$$\Gamma_\xi(\tau) = \lim_{T\to\infty} \frac{1}{2T} \int_{-T}^{T} \xi(t)\xi(t+\tau)\,dt = \frac{1}{2T} \int_{-\infty}^{\infty} \xi_T(t)\xi_T(t+\tau)\,dt \tag{3-55}$$

and the Fourier transform of $\Gamma_\xi(\tau)$ is given by

$$F_\Gamma(\omega) = \int_{-\infty}^{\infty} e^{-j\omega\tau} \left[\frac{1}{2T} \int_{-\infty}^{\infty} \xi_T(t)\xi_T(t+\tau)\,dt \right] d\tau \tag{3-56}$$

Let us extend this expression by $e^{j\omega t}$, and let us include $e^{-j\omega\tau}$ in the inside integral. (It is allowed as this is a constant factor for the inside integral.) Thus

$$F_\Gamma(\omega) = \int_{-\infty}^{\infty} \frac{1}{2T} \int_{-\infty}^{\infty} \xi_T(t)e^{j\omega t}\xi_T(t+\tau)e^{-j\omega(t+\tau)}\,dt\,d\tau \tag{3-57}$$

Again t and $t+\tau$ are symmetrically placed variables, so the sequence of integrations may be exchanged. Thus

$$F_\Gamma(\omega) = \int_{-\infty}^{\infty} \frac{1}{2T} \left[\int_{-\infty}^{\infty} \xi_T(t+\tau)e^{-j\omega(t+\tau)}\,d\tau \right] \xi_T(t)e^{j\omega t}\,dt \tag{3-58}$$

By introducing a new variable $t+\tau=\vartheta$ into the inside integral $d\tau=d\vartheta$, thus the expression

$$\int_{-\infty}^{\infty} \xi_T(\vartheta)e^{-j\omega\vartheta}\,d\vartheta = F_T(\omega) \tag{3-59}$$

is the function of ω only, so

$$F_\Gamma(\omega) = \frac{F_T(\omega)}{2T} \int_{-\infty}^{\infty} \xi_T(t)e^{j\omega t}\,dt = \frac{1}{2T} F_T(\omega) F_T(-\omega) = \frac{|F_T(\omega)|^2}{2T} \quad (3\text{-}60)$$

as $F_T(-\omega) = F_T^*(\omega)$. Let us again cancel our requirement for finite T. Thus

$$F_\Gamma(\omega) = \lim_{T \to \infty} \frac{|F(\omega)|^2}{2T} \quad (3\text{-}61)$$

which only differs from Eq. (3-36) by a factor of 2π. Thus

$$F_\Gamma(\omega) = \int_{-\infty}^{\infty} \Gamma_\xi(\tau)e^{-j\omega\tau}\,d\tau = 2\pi s(\omega) \quad (3\text{-}62)$$

or, using the inverse Fourier transform

$$\boxed{\Gamma_\xi(\tau) = \int_{-\infty}^{\infty} s(\omega)e^{j\omega\tau}\,d\omega} \quad (3\text{-}63)$$

It should be recalled that for a nonperiodic stochastic time function, ω is a continuous variable. The above pair of equations, giving correspondence between the autocorrelation function and the power density, has been introduced by Wiener[15] and Khintchine.[8] In the literature, several forms of these equations are used and below the most frequently used expressions are summarized. $\Gamma_\xi(\tau)$ is a real, even function. Thus

$$\boxed{\begin{aligned} F_\Gamma(\omega) &= \int_{-\infty}^{\infty} \Gamma_\xi(\tau)\cos\omega\tau\,d\tau = 2\int_{0}^{\infty} \Gamma_\xi(\tau)\cos\omega\tau\,d\tau \\ &= 2\pi s(\omega) \quad \text{if} \quad -\infty < \omega < \infty \end{aligned}} \quad (3\text{-}64)$$

$$F_\Gamma(\omega) = \begin{cases} \pi S(\omega) & \text{if} \quad 0 < \omega < \infty & (3\text{-}65) \\ s(f) & \text{if} \quad -\infty < f < \infty & (3\text{-}66) \\ \dfrac{1}{2} S(f) & \text{if} \quad 0 < f < \infty & (3\text{-}67) \end{cases}$$

Knowing one of the power spectrum expressions, it follows from Eq. (3-63) that

$$\Gamma_\xi(\tau) = \overset{\text{double sided}}{\int_{-\infty}^{\infty} s(\omega)\cos\omega\tau\,d\omega} = \overset{\text{single sided}}{\int_{0}^{\infty} S(\omega)\cos\omega\tau\,d\omega} \quad (3\text{-}68)$$

$$\Gamma_\xi(\tau) = \int_{-\infty}^{\infty} s(f)\cos 2\pi f\tau\,df = \int_{0}^{\infty} S(f)\cos 2\pi f\tau\,df \quad (3\text{-}69)$$

Substituting $\tau=0$ into Eq. (3-68), we have

$$\Gamma_\xi(0) = \int_{-\infty}^{\infty} s(\omega)\,d\omega = M[\xi^2(t)] \qquad (3\text{-}70)$$

which is in accordance with Eqs. (3-41) and (3-53).

After the introduction of the autocorrelation function concept, we may now define the gaussian process. Generally, a stochastic process is termed a gaussian process if an ensemble density function follows a multi-dimensional gaussian distribution at any fixed time instant. Most of the stochastic functions in electronics are gaussian, and they have the important property that the distribution type is unchanged after a linear transformation.

In the following, let us assume only two random variables: $\xi_1=\xi_1(t)$ and $\xi_2=\xi_2(t+\tau)$. Their values may be x_1 and x_2. Let the expected value of x_1 and x_2 be zero, their variance σ_x^2, and let them be within the two-variable, normal distribution $f(x_1, x_2)$ defined by Eqs. (2-22) and (2-23). Then

$$\sigma_x^2 = M[\xi^2(t)] = M[\xi^2(t+\tau)] = \int_{-\infty}^{\infty} x^2 f(x)\,dx \qquad (3\text{-}71)$$

$$\Gamma_\xi(\tau) = M[\xi(t)\xi(t+\tau)] = \int\!\!\int_{-\infty}^{\infty} x_1 x_2 f(x_1, x_2)\,dx_1\,dx_2 \qquad (3\text{-}72)$$

The process may be described as gaussian if

$$f(x_1, x_2) = \frac{1}{2\pi\sigma_x^2 \sqrt{1-\varrho}}\, e^{(x_1^2 - 2\varrho x_1 x_2 + x_2^2)/2\sigma_x^2\,(1-\varrho^2)} \qquad (3\text{-}73)$$

where

$$\varrho=\varrho_x(\tau) = \frac{\Gamma_\xi(\tau)}{\sigma_x^2} \qquad (3\text{-}74)$$

This means that the common density function of the instantaneous values always has a normal distribution independent of τ; its parameter ϱ depends on τ.

3-4 TYPICAL AUTOCORRELATION FUNCTIONS AND THEIR POWER SPECTRA

This Section is primarily intended to present illustrative examples in order to appreciate the applications of the autocorrelation function. Also it is proposed to present a few typical properties of electronic noise sources.

In our first problem, let us calculate the autocorrelation function and power spectrum of the most simple deterministic signal

$$u(t) = U = \text{const.} \qquad -\infty < t < \infty \qquad (3\text{-}75)$$

The autocorrelation function is

$$\Gamma(\tau) = \lim_{T \to \infty} \frac{1}{2T} \int_{-T}^{T} u(t)u(t+\tau)\, dt = U^2 \qquad (3\text{-}76)$$

The result is thus the mean square value, as in this case there is a unit corre-spondence between instantaneous values separated by arbitrary time intervals. The power spectrum is calculated from Eq. (3-64). Thus

$$s(\omega) = \frac{1}{\pi} \int_{0}^{\infty} U^2 \cos \omega \tau \, d\tau = \frac{U^2}{\pi} \pi \delta(\omega) \qquad (3\text{-}77)$$

where $\delta(\omega)$ is the Dirac pulse having the properties

$$\delta(\omega) = \begin{cases} 0 & \text{if} & \omega \neq 0 \\ \infty & \text{if} & \omega = 0 \end{cases}$$

$$\int_{-\varepsilon}^{\varepsilon} \delta(\omega)\, d\omega = 1 \quad \text{for any} \quad \varepsilon \quad (\varepsilon > 0); \qquad \delta(-\omega) = \delta(\omega) \quad (3\text{-}78)$$

As a control, let us transform $s(\omega)$ back. Thus

$$\Gamma(\tau) = U^2 \int_{-\infty}^{\infty} \delta(\omega) \cos \omega \tau \, d\omega = \lim_{\varepsilon \to 0} U^2 \int_{-\varepsilon}^{\varepsilon} \frac{\cos \omega \tau}{2\varepsilon}\, d\omega$$

$$= \lim_{\varepsilon \to 0} \left(U^2 \frac{\sin \omega \tau}{2\varepsilon \tau} \right)_{-\varepsilon}^{\varepsilon} = \lim_{\varepsilon \to 0} \left(U^2 \frac{\sin \varepsilon \tau}{\varepsilon \tau} \right) = U^2 \qquad (3\text{-}79)$$

This means that a constant autocorrelation function has a power spectrum of the form $s(\omega) = \text{const.}\ \delta(\omega)$.

In the next problem, the signal

$$u(t) = U \sin(\omega_0 t + \varphi) \qquad (3\text{-}80)$$

will be investigated. Thus

$$\Gamma(\tau) = \lim_{T \to \infty} \frac{U^2}{2T} \int_{-T}^{T} \sin(\omega_0 t + \varphi) \sin(\omega_0 t + \omega_0 \tau + \varphi)\, dt \qquad (3\text{-}81)$$

After elementary trigonometric transformations, we have

$$\Gamma(\tau) = \frac{U^2}{2} \cos \omega_0 \tau \qquad (3\text{-}82)$$

independent of the starting phase φ. This means that Eq. (3-82) is valid for both sine and cosine signals, and meets the conditions of Eqs. (3-48) and (3-49). However, the phase information is lost. The power spectrum, as in Eq. (3-77), is

$$s(\omega) = \frac{U^2}{2\pi} \int_0^\infty \cos \omega_0 \tau \cos \omega \tau \, d\tau$$

$$= \frac{U^2}{2\pi} \left[\int_0^\infty \frac{\cos (\omega - \omega_0) \tau}{2} \, d\tau + \int_0^\infty \frac{\cos (\omega + \omega_0) \tau}{2} \, d\tau \right]$$

$$= \frac{U^2}{2} \left[\frac{\delta(\omega - \omega_0)}{2} + \frac{\delta(\omega + \omega_0)}{2} \right] \qquad (3\text{-}83)$$

This is made up of two half-area Dirac pulses[2,8] placed symmetrical to the origin, disregarding the coefficient $U^2/2$. The single-sided power spectrum is

$$S(\omega) = \frac{U^2}{2} \delta(\omega - \omega_0) \qquad \text{if} \qquad 0 < \omega < \infty \qquad (3\text{-}84)$$

clearly illustrating that the power of a single-frequency deterministic signal is concentrated in a single, well-defined spectrum line.

In the previous two cases, the autocorrelation function and then the power spectrum may easily be determined using known time functions. However, the situation may be the reverse for stochastic signals. In this case $\Gamma(\tau)$ has to be calculated from $S(\omega)$, but the original time function may not be reconstructed because of the lack of phase information.

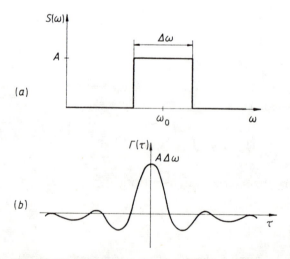

Figure 3-7 Band-limited white noise. (a) Power spectrum.
(b) Autocorrelation function.

Let us now investigate a band-limited white noise having a single-sided power spectrum shown in Fig. 3-7a and given by

$$S(\omega) = \begin{cases} A & \text{if} \quad \omega_0 - \dfrac{\Delta\omega}{2} < \omega < \omega_0 + \dfrac{\Delta\omega}{2} \\ 0 & \text{elsewhere} \end{cases} \tag{3-85}$$

The autocorrelation function is calculated from Eq. (3-68). Thus

$$\Gamma_\xi(\tau) = \int_{\omega_0 - \Delta\omega/2}^{\omega_0 + \Delta\omega/2} A \cos \omega\tau \, d\omega = A\Delta\omega \cos \omega_0\tau \, \frac{\sin \tau\Delta\omega/2}{\tau\Delta\omega/2} \tag{3-86}$$

(see Fig. 3-7b). Let us investigate two limiting cases.

1. Let

$$\tau\Delta\omega \ll 1 \tag{3-87}$$

resulting in

$$\Gamma_\xi(\tau) \cong A\Delta\omega \cos \omega_0\tau \tag{3-88}$$

which closely resembles Eq. (3-82). This suggests that for low bandwidths and short correlation times, even a stochastic signal may be approximated by a sine (cosine) signal of ω_0 angular frequency. In Sec. 2-8 the Rayleigh distribution has been considered and it has been shown that a narrow-band noise may be regarded as a carrier modulated stochastically in both amplitude and phase. It is seen from Eq. (3-86) that the carrier amplitude can be regarded as constant as long as $\Delta\omega$ is small. Figure 3-8 shows that the constant amplitude carrier is well approximated even at $\Delta\omega/2\omega_0=0.1$. In network theory, it is a general rule that the smaller the bandwidth the slower the random phase changes, i.e. the stochastic signal becomes more and more deterministic. In Chapter 5, the noise parameters

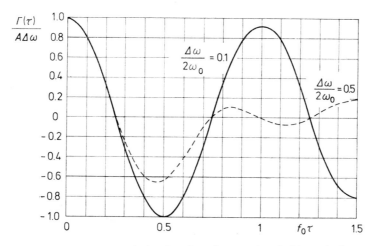

Figure 3-8 Autocorrelation function of narrow-band white noise.[9]
(*Courtesy of Verlag Technik, Berlin.*)

of the linear networks are calculated using this simplifying condition, thus making use of proven methods in classical circuit theory.

2. Let

$$\frac{\Delta\omega}{2} = \omega_0 \tag{3-89}$$

Then, we have

$$\Gamma_{\xi}(\tau) = \int_0^{\Delta\omega} A \cos \omega\tau \, d\omega = A\Delta\omega \frac{\sin \tau\Delta\omega}{\tau\Delta\omega} \tag{3-90}$$

The higher the fixed value of $\Delta\omega$ the faster $|\Gamma_{\xi}(\tau)|$ will tend to zero. For

$$\lim_{\Delta\omega \to \infty} \Gamma_{\xi}(\tau) = A\pi\delta(\tau). \tag{3-91}$$

Thus the autocorrelation time interval of an ideal white noise having infinite bandwidth is zero. This means that there is absolutely no correlation between instantaneous values separated by an infinitely small time (see Figs. 3-1a and c).

Now we can come back to the calculation of the shot-noise power spectrum not detailed in Chapter 1. To this end the model shown in Fig. 1-2 is further

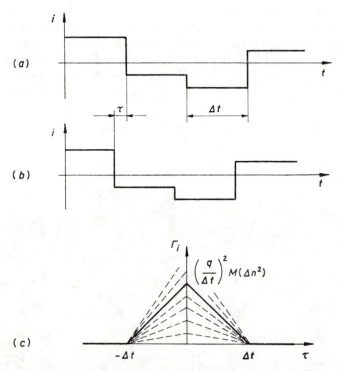

Figure 3-9 Functions involved in the shot-noise model. *(a)* Time function of the current fluctuation $i(t)$. *(b)* Function $i(t-\tau)$. *(c)* Autocorrelation function.

developed. Figure 3-9a illustrates the current fluctuation $i(t)$, i.e. the deviation from the mean value, based on the relation

$$i(t) = \frac{q}{\Delta t}[n(t)-\bar{n}] = \frac{q}{\Delta t}\Delta n(t) \qquad (3\text{-}92)$$

which is a combination of Eqs. (1-7) and (1-8). This has the autocorrelation function

$$\Gamma_i(\tau) = M[i(t)i(t+\tau)] = \lim_{T\to\infty}\frac{1}{2T}\left(\frac{q}{\Delta t}\right)^2\int_{-T}^{T}\Delta n(t)\Delta n(t+\tau)\,dt \qquad (3\text{-}93)$$

according to Eq. (3-47). In order to calculate this function, $i(t)$ has to be shifted either to the left or to the right (see Fig. 3-9b), and the product of the instantaneous values has to be averaged for a long time. As the values

$$\Delta n(t), \Delta n(t+\Delta t), \Delta n(t+2\Delta t), \ldots$$

are independent of each other, the autocorrelation interval may not be longer than $\pm\Delta t$. According to Fig. 3-9c, the time function made up of rectangular current pulses has a convolution resulting in an isosceles triangle. The height of this triangle is different for each pulse, but the average height for a long average time is calculated from Eq. (3-93), by substituting $\tau=0$, to be

$$(q/\Delta t)^2 M(\Delta n^2),$$

i.e.

$$\Gamma_i(\tau) = \begin{cases} \left(\frac{q}{\Delta t}\right)^2 M(\Delta n^2)\left(1-\frac{|\tau|}{\Delta t}\right) & \text{if} \quad |\tau| \leq \Delta t \\ 0 & \text{elsewhere} \end{cases} \qquad (3\text{-}94)$$

As n has a Poisson distribution, we have

$$M(\Delta n^2) = D^2(n) = M(n) \qquad (3\text{-}95)$$

Substituting into Eq. (1-8), we have

$$\Gamma_i(\tau) = \frac{q}{\Delta t}I\left(1-\frac{|\tau|}{\Delta t}\right) \qquad (3\text{-}96)$$

The single-sided power spectrum is calculated from Eq. (3-65), taking into account that the interval for which $\Gamma_i(\tau)$ is defined is $\pm\Delta t$. Thus

$$\boxed{\begin{aligned} S(\omega) &= \frac{2}{\pi}\frac{q}{\Delta t}I\int_0^{\Delta t}\left(1-\frac{\tau}{\Delta t}\right)\cos\omega\tau\,d\tau \\[2mm] &= \frac{2}{\pi}\frac{q}{\Delta t}I\frac{1}{\omega^2\Delta t}(1-\cos\omega\,\Delta t) = \frac{1}{\pi}qI\frac{\sin^2(\omega\Delta t/2)}{(\omega\Delta t/2)^2} \end{aligned}} \qquad (3\text{-}97)$$

For $\omega\Delta t\ll2\pi$ the ratio is approximately unity, so

$$S(\omega) = \frac{1}{\pi}\, qI \tag{3-98}$$

and from Eq. (3-40)

$$S(f) = 2\pi S(\omega) = 2qI \tag{3-99}$$

The mean square value of the fluctuation in a frequency range Δf, according to Eq. (3-41), is

$$M[i^2(t)]_{\Delta f} = \int_f^{f+\Delta f} S(f)\,df = 2qI\Delta f$$

which is the well-known formula for shot noise [see Eq. (1-11)].

The above calculation shows also the limitations of the simple relationship of Eq. (1-11): the intervals Δt may not be decreased arbitrarily. This is explained by the fact that if the charge carriers moving within the electronic device need a finite time interval to overcome a hindrance then the events within this time interval may not be regarded as independent from each other. In many cases, the investigated phenomenon may be modeled not using rectangular pulses as shown in Fig. 3-9 but using pulses of another shape. In these cases, the multiplying factor of $S(\omega)$ also differs from $(\sin x/x)^2$.

In the above example, the investigated time interval has been divided into arbitrary parts, and the pulses have been obtained by this arbitrary subdivision. However, there are types of electronic noise that originally have pulse waveforms (burst noise and some processes connected with voltage breakdown). The time function of these processes is illustrated by the so-called random telegraph signal shown in Fig. 3-10.

For simplicity, let us investigate a symmetrical random telegraph signal, the values $+U$ and $-U$ having equal probabilities of 0.5. Let the changeovers follow a Poisson distribution; this means that if there are v changeovers during unit time, then the probability of having $0, 1, 2, ..., n$ changeovers during a time interval τ will be

$$P(\eta = n) = \frac{(v\tau)^n}{n!}\, e^{-v\tau} \tag{3-100}$$

It should be noted that in this case the time intervals between changeovers follow an exponential distribution with a density function according to Eq. (2-144).

As the process is assumed to be stationary, the product $\xi(t)\xi(t+\tau)$ may only be $+U^2$ or $-U^2$, depending on whether the number of changeovers during a time-interval τ is $0, 2, 4, ...$ or $1, 3, 5, ...$. Thus after averaging we have

$$M[\xi(t)\xi(t+\tau)] = U^2\times(\text{the probability of even changeovers}) -$$
$$- U^2\times(\text{the probability of odd changeovers}) \tag{3-101}$$

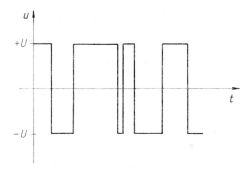

Figure 3-10 Random telegraph signal.

Combining Eqs. (3-100) and (3-101) gives

$$M[\xi(t)\xi(t+\tau)] = U^2 e^{-v\tau}\left[1+\frac{(v\tau)^2}{2!}+\frac{(v\tau)^4}{4!}+\dots\right] - U^2 e^{-v\tau}\left[\frac{v\tau}{1!}+\frac{(v\tau)^3}{3!}+\dots\right]$$

(3-102)

Rearrangement of this expression yields

$$\Gamma_\xi(\tau) = U^2 e^{-2v\tau}$$

(3-103)

Note that $\Gamma_\xi(\tau)$ is always even, therefore the correct end result is

$$\Gamma_\xi(\tau) = U^2 e^{-2v|\tau|}$$

(3-104)

According to Appendix B, the Fourier transform of this expression is

$$s(\omega) = \frac{4U^2 v}{4v^2+\omega^2}$$

(3-105)

This is called a Lorentz spectrum.

3-5 RESPONSE OF A LINEAR NETWORK DRIVEN BY A STOCHASTIC SIGNAL[1, 2, 3, 7, 9, 10, 12]

For small signals, electronic devices behave as linear networks. Several noise sources may be present in a single device such as a transistor, and between the noise sources and output terminals there are usually resistive and reactive network elements of the equivalent circuit. This network has an effect on the output autocorrelation function and power spectrum.

A linear time-invariant network may be characterized by its weight function $h(\vartheta)$. If the input driving signal is $x(t)$, the output response is

$$y(t) = \int_{-\infty}^{\infty} h(\vartheta)x(t-\vartheta)\,d\vartheta$$

(3-106)

In the case of physically realizable networks, $h(\vartheta)=0$ if $\vartheta<0$. The stability criterion is

$$\int_{-\infty}^{\infty} |h(\vartheta)|\, d\vartheta < \infty \qquad (3\text{-}107)$$

Equation (3-106) is really an expression of the system's "memory": the response function is formed by multiplying the infinitely narrow driving elements falling in the interval $(t-\vartheta, t)$ by the weight function $h(\vartheta)$.

Let us assume a linear invariant network driven by an ideal white noise with independent instantaneous values separated by an arbitrarily short time interval $[\Gamma_{\xi}(\tau)=0$ if $|\tau|>0]$. However, the response function will include not only the driving signal value pertaining to the given instant t but also earlier signal values, although with decreased weight, unless the weight function $h(\vartheta)$ shows a very fast convergence to zero with increasing ϑ. This means that the autocorrelation functions of the driving and response waveforms will be different.

Let us assume an input driving signal $\xi(t)$ which is stationary and ergodic with an autocorrelation function $\Gamma_{\xi}(\tau)$, and let us determine the autocorrelation function $\Gamma_{\eta}(\tau)$ of the response $\eta(t)$. The product $\eta(t)\eta(t+\tau)$ is first calculated. Thus

$$\eta(t)\eta(t+\tau) = \int_{-\infty}^{\infty} h(\vartheta)\xi(t-\vartheta)\, d\vartheta \int_{-\infty}^{\infty} h(\Theta)\xi(t+\tau-\Theta)\, d\Theta$$

$$= \iint_{-\infty}^{\infty} h(\vartheta)h(\Theta)\xi(t-\vartheta)\xi(t+\tau-\Theta)\, d\Theta\, d\vartheta \qquad (3\text{-}108)$$

Here two kinds of integration variables ϑ and Θ have been introduced for formal reasons only. According to Eq. (3-2)

$$\Gamma_{\eta}(\tau) = \lim_{T\to\infty} \frac{1}{2T} \int_{-T}^{T} \eta(t)\eta(t+\tau)\, dt \qquad (3\text{-}109)$$

Thus Eq. (3-108) should be averaged in time. As t, ϑ and Θ are independent of each other, the sequence of integrations may again be changed. Let us first calculate Eq. (3-109). Thus

$$\Gamma_{\eta}(\tau) = \lim_{T\to\infty} \iint_{-\infty}^{\infty} h(\vartheta)h(\Theta)\left[\frac{1}{2T}\int_{-T}^{T}\xi(t-\vartheta)\xi(t+\tau-\Theta)\, dt\right] d\Theta\, d\vartheta$$

$$= \iint_{-\infty}^{\infty} h(\vartheta)h(\Theta)\Gamma_{\xi}(\tau-\Theta+\vartheta)\, d\Theta\, d\vartheta \qquad (3\text{-}110)$$

The application of this expression is rather awkward. However, the following relationships are known:

$$h(t) \xleftarrow{\text{Fourier transform}} H(\omega)$$

$$\Gamma(\tau) \xleftarrow{\hspace{2cm}} s(\omega) \qquad (3\text{-}111)$$

where $H(\omega)$ is the transfer function of the network. Utilizing these, the Fourier transform of $\Gamma_\eta(\tau)$ is given by

$$s_\eta(\omega) = \frac{1}{2\pi} \int_{-\infty}^{\infty} \Gamma_\eta(\tau) e^{-j\omega t}\, d\tau$$

$$= \frac{1}{2\pi} \int_{-\infty}^{\infty} e^{-j\omega\tau} \left[\iint_{-\infty}^{\infty} h(\vartheta) h(\Theta) \Gamma_\xi(\tau - \Theta + \vartheta)\, d\Theta\, d\vartheta \right] d\tau \qquad (3\text{-}112)$$

Let us multiply by $e^{j\omega(\vartheta - \Theta)}$ and transform the triple integral into a product of three integrals (the independence of the three variables allows this). Thus

$$s_\eta(\omega) = \int_{-\infty}^{\infty} h(\vartheta) e^{j\omega\vartheta}\, d\vartheta \int_{-\infty}^{\infty} h(\Theta) e^{-j\omega\Theta}\, d\Theta \frac{1}{2\pi} \int_{-\infty}^{\infty} \Gamma_\xi(\tau - \Theta - \vartheta) e^{-j\omega(\tau - \Theta + \vartheta)}\, d\tau$$

$$(3\text{-}113)$$

As $h(\vartheta)$ is real, the first two Fourier transforms form a conjugate pair [similar to Eq. (3-26)], so their product is the square of the absolute value of the transfer function. The third integral being just $s_\xi(\omega)$ we have

$$\boxed{s_\eta(\omega) = |H(\omega)|^2 s_\xi(\omega)} \qquad (3\text{-}114)$$

This relation, comprising only frequency domain parameters, is simple and instructive. It is normally used instead of Eq. (3-110) for the calculation of the auto-correlation functions, according to the scheme

$$\Gamma_\xi(\tau) \xrightarrow{\mathscr{F}} s_\xi(\omega) |H(\omega)|^2 \xrightarrow{\mathscr{F}^{-1}} \Gamma_\eta(\tau)$$
$$\mathscr{F} \uparrow \qquad\qquad\qquad (3\text{-}115)$$
$$h(\vartheta)$$

The transformation $h(\vartheta) \to H(\omega)$ is frequently not performed separately as the transfer function of the network is normally known.

As an example, let us calculate the autocorrelation function of the output signal for a low-pass filter with a transfer function

$$H(\omega) = \frac{1}{1 + j(\omega/\omega_a)} \qquad (3\text{-}116)$$

when the input is driven by white noise. Let the single-sided power spectrum of the white noise again be

$$S_\xi(\omega) = A \qquad (3\text{-}117)$$

At the filter output we have

$$S_\eta(\omega) = \frac{A}{1 + (\omega/\omega_a)^2} \qquad (3\text{-}118)$$

from which, according to Eq. (3-68) (see Appendix B), it follows that

$$\Gamma_n(\tau) = \int\limits_0^\infty \frac{A\cos\omega\tau}{1+(\omega/\omega_a)^2}\,d\omega = A\omega_a\frac{\pi}{2}\,e^{-|\omega_a\tau|} \tag{3-119}$$

The power spectra and the autocorrelation functions are shown in Fig. 3-11.

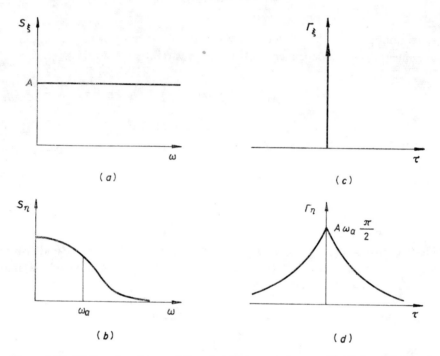

Figure 3-11 White-noise characteristics. *(a)* Power spectrum, without band-limiting. *(b)* Power spectrum, with band-limiting. *(c)* Autocorrelation function, without band-limiting. *(d)* Autocorrelation function, with band-limiting.

In the case of an input signal with normal distribution, the output signal will also have normal distribution. A band-limiting network driven by an input with anormal distribution will provide a response that has a distribution closer to normal because, according to Eq. (3-106), $y(t)$ is a function of not a single instantaneous value but of an infinite number of instantaneous values weighted by $h(\vartheta)$. Regarding this series of instantaneous values as members of a sum distribution, a normal distribution should be attained as a consequence of the central limit theorem. The actual situation is much more complicated: the instantaneous values of the input signal (roughly inside the autocorrelation interval τ_k) are not independent from each other, so the central limit theorem may not be applied. As the problem at present has no general solution,[13,14] the only qualitative statement possible is that the higher the time constant of the band-limiting network compared with the autocorrelation interval of the input signal the better the output signal distribution will approximate the normal distribution.

PROBLEM

3-1 Calculate the autocorrelation function and power spectrum of the triangnlar pulse shown in Fig. 3-6.

Answer:

$$\xi(t) = X\frac{t}{t_1} \qquad \text{if} \qquad 0 < t < t_1$$

$$\xi(t-\tau) = X\frac{t-\tau}{t_1} \qquad \text{if} \qquad \tau < t < t_1 + \tau$$

The autocorrelation function (see Fig. 3-6b) is

$$\Gamma_\xi(\tau) = \Gamma_\xi(-\tau) = M\left[X\frac{t}{t_1}X\frac{t-\tau}{t_1}\right] = \frac{X^2}{3}\left[1 - \frac{3}{2}\frac{|\tau|}{t_1} + \frac{1}{2}\frac{|\tau|^3}{t_1^3}\right] \qquad 0 < |\tau| < t_1$$

The power spectrum, according to Eq. (3-64), is

$$S(\omega) = \frac{1}{\pi}\frac{X^2}{3}\int_0^{t_1}\left[1 - \frac{3}{2}\frac{\tau}{t_1} + \frac{1}{2}\left(\frac{\tau}{t_1}\right)^3\right]\cos\omega\tau\,d\tau$$

Partial integration and trigonometrical rearrangements provide the result

$$S(\omega) = \frac{t_1}{2\pi}\frac{X^2}{(\omega t_1)^4}[(\omega t_1)^2 + 2(1 - \cos\omega t_1 - \omega t_1\sin\omega t_1)]$$

REFERENCES

[1] Bendat, J. S.: *Principles and Applications of Random Noise Theory,* John Wiley, New York, 1958.

[2] — and A. G. Piersol: *Measurement and Analysis of Random Data,* John Wiley, New York, 1966.

[3] — and A. G. Piersol: *Random Data: Analysis and Measurement Procedures,* Wiley Interscience, New York, 1971.

[4] Bennet, W. R.: "Methods for Solving Noise Problems", *Proc. IRE*, vol. 44, no. 5, pp. 609—637, 1956.

[5] Bracewell, R.: *The Fourier Transform and Its Applications*, McGraw-Hill, New York, 1965.

[6] Champeney, D. C.: *Fourier Transforms and Their Physical Applications*, Academic Press, New York, 1973.

[7] Fodor, G.: *The Analysis of Linear Systems* (Lineáris rendszerek analízise), Műszaki Könyv-kiadó, Budapest, 1967 (in Hungarian).

[8] Khintchine, A.: "Correlation Theory of the Stationary Stochastic Process", *Math. Ann.*, vol. 109, no. 4, pp. 604—615, 1934.

[9] Lange, F. H.: *Korrelationselektronik*, Verlag Technik, Berlin, 1962.

[10] Mix, D. F.: *Random Signal Analysis*, Addison-Wesley, Reading, 1969.

[11] Papoulis, A.: *Probability, Random Variables and Stochastic Processes*, McGraw-Hill, New York, 1965.

[12] Roth, P. R.: "Effective Measurements Using Digital Signal Analysis", *IEEE Spectrum*, vol. 8, no. 4, pp. 62—70, 1971.

[13] Sutcliffe, H.: "Pseudo-Random Noise in the Time and Frequency Domains", *Period. Poly-techn.*, vol. 15, no. 1, pp. 63—70, 1971.

[14] Tomlinson, G. H. and P. Galvin: "Design Criterion for Generation of Gaussian Signals from Smoothed m-Sequences", *Electr. Lett.*, vol. 12, no. 14, pp. 349—350, 1976.

[15] Wiener, N.: "Generalized Harmonic Analysis", *Acta Mathematica*, vol. 55, pp. 117—258, 1930.

FOUR

THE PHYSICAL ASPECTS OF NOISE

The utilization of the concepts in probability theory described in previous Chapters results in models that provide a good approximation of most of the noise phenomena in electronic devices. However, these models have been constructed by neglecting several parameters, and while this is permitted for a general description a more rigorous treatment is necessary for the analysis of physical aspects. The relation $\overline{i^2}=2qI\Delta f$ for shot noise given in Chapter 1 [see Eq. (1-11)] has been further developed in Chapter 3. From Eq. (3-97) we have

$$\overline{i^2} = 2qI\frac{\sin^2(\omega\Delta t/2)}{(\omega\Delta t/2)^2}\Delta f \qquad (4\text{-}1)$$

The multiplying factor $(\sin x/x)^2 \leqq 1$ means a finite upper-frequency limit that is evidently connected with the geometrical layout and physical operation of the noise source. However, expression (4-1) is only valid for the current fluctuation model shown in Fig. 1-2. In an actual device, the stochastic processes are made up of elementary pulses having different shapes.

4-1 SHOT NOISE

Shot noise is generated when the charge carriers supplying the current pass through an energy gap. The energy gaps within electronic devices are always brought about by some heterogeneous structure: in a vacuum tube by the cathode metal and the vacuum, while in a solid state device by two layers with different conduction types or two different solid materials.[118] On the other hand, the dimensions of the vacuum space or of the solid state device depletion layers cannot be arbitrarily decreased, both for technological and theoretical reasons. The transit process of the charge carriers is concluded by the passage through the energy gap and the finite space involving this gap, thus requiring a finite time interval. Therefore our previous models assuming Dirac pulses are merely approximate ones.

Figure 4-1 shows the cross section of a vacuum diode with plane electrodes. The electron with a charge q has zero initial velocity at the cathode and travels to the anode with constant acceleration. The transit time (time of flight) is

$$t_f = \frac{2d}{v_{max}} \tag{4-2}$$

where v_{max} is the maximum velocity at the anode. If, in the meantime, no further electron is emitted from the cathode, the external circuit has also to pass a charge q during time interval t_f. However, the time dependence of the current flowing in the external circuit is not yet known.

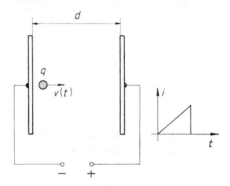

Figure 4-1 Movement of an electron
in a vacuum between planar electrodes.

When moving along a path length dx, the additional energy gained by the electron is

$$dW = F\,dx = qE\,dx = q\frac{U_{ac}}{d}\,dx \tag{4-3}$$

This additional energy is supplied by the power supply:

$$dW = U_{ac}\,i(t)\,dt \tag{4-4}$$

where $i(t)$ is the instantaneous value of the current flowing in the external circuit. Using Eqs. (4-3) and (4-4), we have

$$i(t) = \frac{q}{d}\frac{dx}{dt} = \frac{q}{d}v(t) \tag{4-5}$$

In the case of constant acceleration, $v(t)$ increases linearly up to v_{max}:

$$i(t) = \frac{q}{d}v_{max}\left(\frac{t}{t_f}\right) = \frac{2q}{t_f}\left(\frac{t}{t_f}\right) \tag{4-6}$$

where Eq. (4-2) has been utilized to eliminate d. Let the number of electrons emitted simultaneously be n. This number is a random variable with Poisson distribution having an expected value of $M(n)$. The mean value of the current is

$$I = M[ni(t)] = M(n)\frac{2q}{t_f^2}\int_0^{t_f}\left(\frac{t}{t_f}\right)dt = M(n)\frac{q}{t_f} \qquad (4\text{-}7)$$

which is directly self-evident. The autocorrelation function of the current pulses $ni(t)$ is

$$\Gamma_i = M[ni(t)ni(t+\tau)] \qquad (4\text{-}8)$$

As $M[ni(t)]\neq0$, let us convert expression (4-8) according to Eq. (3-45). Thus

$$\Gamma_i = M[\Delta ni(t)\Delta ni(t+\tau)]+I^2 = \Gamma_i + I^2 \qquad (4\text{-}9)$$

where $\Delta n = n - M(n)$, and Γ_i is the autocorrelation function of the fluctuation component. In the following, the investigation of Γ_i will be sufficient:

$$\Gamma_i = M(\Delta n^2)\frac{4q^2}{t_f^2}M\left[\frac{t}{t_f}\frac{t+\tau}{t_f}\right] = \frac{4qI}{t_f}\left[1 - \frac{3}{2}\frac{|\tau|}{t_f}+\frac{1}{2}\frac{|\tau|^3}{t_f^3}\right] \qquad (4\text{-}10)$$

where the relation (2-89), $D^2(n)=M(\Delta n^2)=M(n)$, which pertains to the Poisson distribution, has been used. Equation (4-7) and the example given in Problem 3-1 have also been utilized.

The power spectrum[95] is given by

$$s(\omega) = \frac{1}{2\pi}qI\frac{4}{(\omega t_f)^4}[(\omega t_f)^2+2(1-\cos\omega t_f-\omega t_f\sin\omega t_f)] \qquad (4\text{-}11)$$

which is difficult to evaluate in this form. Expanding the angular functions for $\omega t_f \ll 1$ and simplifying, we have

$$s(\omega) = \frac{1}{2\pi}qI\left[1 - \frac{(\omega t_f)^2}{18}+ \ldots\right] \qquad (4\text{-}12)$$

Expressing now the single-sided density function $S(f)$ and taking into account Eqs. (3-64) to (3-67), according to which $S(f)=4\pi s(\omega)$,

$$\overline{i^2} = S(f)\,\Delta f = 2qI\left[1 - \frac{(\omega t_f)^2}{18}+ \ldots\right]\Delta f = 2qIF_{fp}^2\,\Delta f \qquad (4\text{-}13)$$

where $F_{fp}\leq1$ which is a factor depending on the transit angle ωt_f for a planar diode.

For other geometrical configurations and field distributions both the form of the current pulses and the power spectrum will be modified. For a cylindrical diode, we have instead of Eq. (4-3)

$$dW = F\,dx = qE(r)\,dr = \frac{qU_{ac}}{r\ln \varrho_a}\,dr \qquad (4\text{-}14)$$

$$\varrho_a = \frac{r_a}{r_c}$$

thus, substituting Eq. (4-4) and $dr/dt = v(t)$,

$$i(t) = \frac{q}{\ln \varrho_a}\frac{v(t)}{r(t)} \qquad (4\text{-}15)$$

The function $i(t)$ may be plotted utilizing the relations

$$v(r) = v_{max}\sqrt{\frac{\ln \varrho}{\ln \varrho_a}} \qquad \varrho = \frac{r}{r_c} \qquad (4\text{-}16)$$

$$t(r) = \int_{r_c}^{r}\frac{dr}{v(r)} = \frac{2r_c}{v_{max}}\sqrt{\ln \varrho_a}\int_{0}^{\sqrt{\ln \varrho}} e^{y^2}\,dy \qquad (4\text{-}17)$$

where

$$\ln \varrho = \ln \frac{r}{r_c} = y^2 \qquad dr = 2r_c\,ye^{y^2}$$

In Fig. 4-2, $i(t)/I$ is plotted for three diodes having different electrode configurations.[35, 105] If $r_a/r_c \to \infty$ then $i(t)$ will approximate the Dirac pulse, so the Fourier transform of the autocorrelation function would result, according

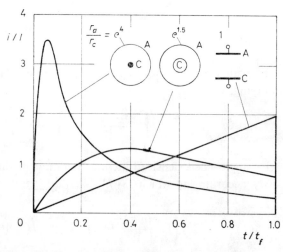

Figure 4-2 Shape of current pulse generated in the external circuit for different electrode configurations. (*Courtesy of Siemens.*)

to Eq. (3-91), in a white power spectrum. However, for practically realizable structures, taking into account the factor F_{fc}^2 depending on the transit time, we have, according to Spenke,[103]

$$F_{fc}^2 = \left[\frac{1}{z_a} \int_0^{z_a} \cos \omega t_f \frac{\psi(\sqrt{z})}{\psi(\sqrt{z_a})} \, dz \right]^2 + \left[\frac{1}{z_a} \int_0^{z_a} \sin \omega t_f \frac{\psi(\sqrt{z})}{\psi(\sqrt{z_a})} \, dz \right]^2 \quad (4\text{-}18)$$

where $z = \ln \varrho$ and

$$\psi(\sqrt{z}) = \int_0^{u=\sqrt{z}} e^{u^2} \, du \quad (4\text{-}19)$$

Figure 4-3 shows the factors F_f in Eqs. (4-11) and (4-18). The three curves nearly coincide in the interval $0 < \omega t_f < \pi/2$, so the two-term approximation given in Eq. (4-13) may be applied for F_f in all three cases. The transit time is also dependent on the structure. If $r_a/r_c \to \infty$ then $(t_f)_c \to \frac{1}{2}(t_f)_p$ (assuming the same anode voltages and electrode distances).[87]

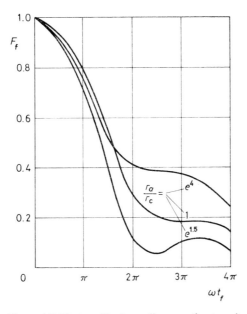

Figure 4-3 Factor F_f depending on the transit angle for different electrode configurations. (*Courtesy of Siemens.*)

In the depletion layers of a solid state device p–n junction, the velocity of the charge carriers is nearly constant. If the external circuit is a short circuit (i.e. it does not comprise delaying elements) then the current pulse due to the passing of a single charge carrier will be rectangular. The power spectrum, according to Eq. (3-97), is

$$S(f) = 2qI\left(\frac{\sin \omega t_f/2}{\omega t_f/2}\right)^2 = 2qI\left[1 - \frac{(\omega t_f)^2}{12} + \dots\right] \qquad (4\text{-}20)$$

At low frequencies, $S(f)$ is constant and independent of the pulse shape. If $f = 2/\pi t_f$, then the power density will be 5 to 10 percent less compared with the low-frequency value, according to the different F_f values. This means that the spectrum of the shot noise may be regarded as white in a wide frequency range.

The frequency dependence of the shot-noise power density is affected by other factors in addition to the finite transit time. This is explained by the fact that usually parasitic elements, such as electrode capacitance, lead resistance, and inductance, are present between the noise source and the external circuit, and these form a low-pass network. According to Sec. 3-5, this modifies both the power spectrum and the autocorrelation function. For instance, if the noise current does not flow directly into the external circuit but through an RC circuit made up of the series resistance R and electrode capacitance C, then the power spectrum, according to Eq. (3-114), is

$$\boxed{S'(f) = 2qI\frac{F_f^2}{1 + (2\pi fRC)^2}} \qquad (4\text{-}21)$$

where $[1 + (2\pi fRC)^2]^{-1} = |H|^2$ is the power transfer function of the network. In practical cases, the effect of the denominator in Eq. (4-21) is frequently much higher than the effect of the numerator.

In our previous calculations, the appearance of the charge carriers has been regarded as an independent event. For most of the solid state devices and for a vacuum diode operating in the saturation or exponential region, this is explained by the following reasoning:

1. The starting current of a vacuum diode comprises few electrons with nearly independent fields.
2. The saturation current is provided by fast-moving electrons. According to the relation $\varrho = j/v$, a small space charge is generated by a given current.
3. In solid state devices, the charge providing the current is balanced by the fixed charge of the ion grid, by the charge of the majority carriers and by the surface charges. As a consequence of charge neutrality, there is no space charge.

However, in the vicinity of the cathode in a vacuum tube operating in the space charge region, the electrons are slowed down and form a space charge cloud. This affects the electrons that are subsequently emitted. The space charge cloud smoothes the fluctuations of the cathode current, so

$$\overline{i^2} = 2qIF_s^2\, \Delta f \qquad (4\text{-}22)$$

Here $F_s < 1$ is the noise attenuation factor characteristic for the space charge operating mode.[119] Thus

$$F_s^2 = \frac{9(1-\pi/4)kT_c}{q(U_a-U_{min})} \qquad (4\text{-}23)$$

where $k = 1.38 \times 10^{-23}$ J/K is the Boltzmann constant, T_c is the cathode temperature, U_a is the anode voltage (in the case of a tube with grids, the effective voltage in the plane of the control grid), and U_{min} is the potential minimum in the vicinity of the cathode.

4-2 PARTITION NOISE

In many cases, the charge carriers emitted within an electronic device follow random paths. Let us assume that according to the general arrangement shown in Fig. 4-4a a constant number of charge carriers are emitted per unit time from electrode 0, and these charge carriers reach either electrode 1 or electrode 2. Let the probability of reaching electrode 1 be p_1 and the probability of reaching electrode 2 be $p_2 = 1 - p_1$. (In this case, no charge carriers will return to the emitting electrode.) This is again a simple alternative, so the model presented in Sec. 1-1 may be utilized with slight modification.

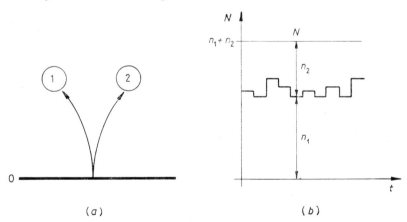

(a)   (b)

Figure 4-4 (a) Current partition between two collector electrodes.
(b) Model for the fluctuation of current components.

Suppose that the number of evenly emitted charges during time Δt is Nq, where N is a constant, i.e. it is not a random variable. The average value of the total current is then

$$I = \bar{i} = \frac{Nq}{\Delta t} \qquad (4\text{-}24)$$

which is randomly distributed in two parts

$$i_1 = \frac{n_1 q}{\Delta t} \quad \text{and} \quad i_2 = \frac{n_2 q}{\Delta t} \tag{4-25}$$

where i_1 and i_2 are the instantaneous values of the currents provided by n_1 and n_2 electrons, respectively. n_1 and n_2 have binomial distributions, so according to Eqs. (2-81) and (2-82), we have

$$M(n_1) = N p_1 \tag{4-26}$$

and

$$D^2(n_1) = N p_1 p_2 = D^2(n_2) \tag{4-27}$$

The expression $D^2(n_1) = D^2(n_2)$ is not only a formal consequence of Eq. (4-27). N being constant in any interval Δt, the fluctuation in n_1 creates a fluctuation in n_2 of equal magnitude but of opposite sense (see Fig. 4-4b). Thus

$$D^2(i_1) = D^2(i_2) = \left(\frac{q}{\Delta t}\right)^2 N p_1 (1 - p_1) \tag{4-28}$$

Combining Eqs. (4-25) and (4-26), we have

$$I_1 = M(i_1) = \frac{q}{\Delta t} M(n_1) = \frac{q}{\Delta t} N p_1 \tag{4-29}$$

Furthermore, it is easily shown that

$$1 - p_1 = \frac{I_2}{I} = \frac{I_2}{I_1 + I_2} \tag{4-30}$$

Consequently

$$D^2(i_1) = \frac{q}{\Delta t} \frac{I_1 I_2}{I_1 + I_2} \tag{4-31}$$

The autocorrelation function and power spectrum of the stochastic process shown in Fig. 4-4b have already been calculated in Sec. 3-4; the single-sided power spectrum is given, as in Eq. (3-99), by

$$\boxed{S(f) = 2q \frac{I_1 I_2}{I_1 + I_2} = 2q(I_1 \| I_2)} \tag{4-32}$$

For $I_1 \ll I_2$,

$$S(f) \cong 2q I_1 \tag{4-33}$$

Thus, the analogy between the partition noise and shot noise is clearly seen. If the condition of Eq. (4-33) is met, the instantaneous values follow a Poisson distribution; otherwise, we have a binomial distribution. In practical cases, N is normally so high that both may be approximated by a normal distribution (see Sec. 2-7).

It is well known that within an electronic device the emission of electrons is not even and this fluctuation brings about the shot noise. Taking into account this fluctuation too, Eq. (4-32) may be rewritten in the following form, referred to electrode 1:[20,129]

$$S_{i1}(f) = \left(\frac{I_1}{I}\right)^2 2qI + 2q\frac{I_1 I_2}{I} \qquad (4\text{-}34)$$

This means that now $S_{i1}(f) \neq S_{i2}(f)$; the latter may be calculated in a similar way to Eq. (4-34) by incorporating I_2 into the numerator of the first term.

4-3 THERMAL NOISE

In homogeneous conductors or semiconductors, there are no energy pedestals, but in spite of this, noise is generated even without current flow. This is due to the thermal motion of charge carriers and to the nonideal structure of the conductor or semiconductor. (This latter prevents the achievement of infinite conductance.) It may be proven generally[142] that there is direct proportionality between the conductance of solids and the root mean square value of the noise current generated within these solids. Our corpuscular kinetic model is far from being general. Nevertheless, it is used because it allows direct conclusions concerning the microphysical processes within the solids.[36] This is evident from the following model of current conduction which will be needed to establish our subsequent results.

Assume a solid of ideal structure: the placement of the atoms forming the crystal lattice is strictly periodic. According to quantum mechanics, this kind of structure allows free passage of an electron or a lack of an electron—a hole.[9] However, the periodicity may be disturbed in several ways, such as the presence of crystal boundaries in polycrystalline materials, crystal defects, and contaminating atoms in single crystal materials, as well as by the thermal motion of the atoms forming the grid, which is a basic effect. There is an inelastic impact between the charge carriers and these defects of the solid. The average energy of the charge carriers after impact is proportional to the temperature. If a field is generated within the conductor, the additional energy of the charge carriers accelerated by this field is lost during each impact, so externally applied energy is needed to enable their further passage.

Let us assume that in the instant $t=0$ the number of charge carriers that suffer impacts is C, and let us investigate their subsequent paths. The impacts being independent events, the change in the number of charge carriers that have not yet suffered impacts between t and $t+\Delta t$ is

$$dC(t) = \frac{C(t)\,dt}{\Theta} \qquad (4\text{-}35)$$

where Θ is the average time between impacts (the change is in a decreasing sense).

Solving this differential equation, the density function of the time distribution between two impacts may be determined. Thus

$$f(t) = \frac{1}{\Theta} e^{-t/\Theta} \qquad t > 0 \tag{4-36}$$

This is in agreement with the density function (2-144) which has the first- and second-order moments $M(t) = \Theta$ and $M(t^2) = 2\Theta^2$, respectively [see Eq. (2-147)]. Let the field strength in a prism-shaped solid be E_x. The path of one of the electrons may have the shape shown in Fig. 4-5 because of the impacts. The projection of these path elements in direction x is given by

$$x_i = \frac{a_x}{2} t_i^2 \tag{4-37}$$

where t_i is the time interval from the $(i-1)$-th impact to the i-th impact and the acceleration of the electron is

$$a_x = -\frac{q}{m^*} E_x \tag{4-38}$$

where m^* is the effective mass of the electron in the crystal. Taking into account that according to Eq. (2-147) $M(t^2)/2 = \Theta^2$, the distance covered during a large number of impacts K will be given by

$$x_K = \frac{a}{2} \sum_1^K t_i^2 = \frac{a}{2} KM(t^2) = aK\Theta^2 \tag{4-39}$$

and the time needed to cover this distance is

$$t_K = \sum_j^K t_i = KM(t) = K\Theta \tag{4-40}$$

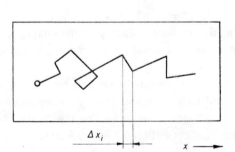

Figure 4-5 Path of an electron moving in an electric field within a solid. The relatively slow movement in the x-direction is superimposed on a random thermal movement.

From these expressions, the average drift velocity is

$$\bar{v}_d = \frac{x_K}{t_K} = a\Theta = -\frac{q}{m^*}E_x\Theta \tag{4-41}$$

If the number of electrons moving with a velocity v_d within a unit volume is n_0 then the current passing through a unit area is

$$J_x = (-q)n_0\bar{v}_d = \frac{q^2}{m^*}n_0\Theta E_x \tag{4-42}$$

Utilizing the definition of conductivity, we have

$$\gamma = \frac{J_x}{E_x} = \frac{q^2}{m^*}n_0\Theta \tag{4-43}$$

so γ is directly proportional to the average time interval Θ between two impacts. The order of Θ is 10^{-14} to 10^{-13} s, and the average free distance is between 10 and 30 nm.

During the calculation of Eq. (4-43), our model has been basically classic, in spite of the fact that the results of quantum mechanics have been taken into account through the introduction of m^*. The same result can be achieved using more accurate quantum mechanical calculations.[64]

For the case of $E=0$, there is no macroscopically measurable current flow within the solid. However, there is a random thermal motion of the charge carriers which has a velocity of v which is much higher than the practically occurring values of v_d. The direction of the charge carrier motion is changed during each impact so there are microscopic current changes both in the volume in question and in the external circuit connected to it. The average value of the velocity is not changed significantly by impact as the sites of impact are at the same temperature.

Let the number of charge carriers having equal polarity again be n_0 within the unit volume of the solid with cross section A, as shown in Fig. 4-6. However,

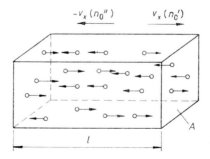

Figure 4-6 The x-direction component of the thermal velocity of charge carriers.

the velocity components in the x directions should now be different: $+v_x$ for n_0' charge carriers, and $-v_x$ for $n_0'' = n_0 - n_0'$ charge carriers. In this case, the current flowing in one direction is

$$i' = AJ' = An_0'qv_x = \frac{N'}{l} qv_x \qquad (4\text{-}44)$$

and the current flowing in the opposite direction is given by

$$i'' = An_0''q(-v_x) = -\frac{N-N'}{l} qv_x \qquad (4\text{-}45)$$

where $N = Aln_0$, $N' = Aln_0'$, $N'' = Aln_0''$ is the number of charge carriers within the entire volume. The difference between the two currents is

$$i(t) = i' - i'' = \frac{qv_x}{l}(2N' - N) = \frac{2qv_x}{l}\left[N'(t) - \frac{N}{2}\right] \qquad (4\text{-}46)$$

The sign of Eq. (4-46) depends on the type of charge carriers and the instantaneous value of N'. The sign may be disregarded below as only the mean square value of i will be needed; our calculation will thus be valid both for electrons and holes.

If the two kinds of charge carriers are simultaneously present, N' should be calculated by taking into account the holes of velocity $+v_x$ and electrons of velocity $-v_x$, and N'' should be calculated by taking into account electrons and holes moving in opposite directions.

N' and N'' are random variables pertaining to mutually exclusive events, their expected value being $N/2$ and their distribution being binomial. According to Eq. (2-84) we have

$$D^2(N') = \frac{M(N')}{2} = \frac{N'}{4} \qquad (4\text{-}47)$$

The current fluctuation $i(t)$ is the consequence of random collisions of carriers and lattice imperfections. The distribution of time intervals between collisions being exponential [see Eq. (4-36)], the autocorrelation function of $i(t)$ is

$$\Gamma_i(\tau) = M[i(t)i(t+\tau)] = \frac{4q^2\overline{v_x^2}}{l^2} D^2(N')e^{-\tau/\Theta} = \frac{q^2\overline{v_x^2}}{l} n_0 A e^{-\tau/\Theta} \qquad (4\text{-}48)$$

which is explained by the fact that from the charge carriers moving together in time instant t only the fraction $e^{-\tau/\Theta}$ will not have collided after the time interval τ. From the expression (3-67) of the Wiener–Khinchine theorem (Appendix B), we have

$$S_i(f) = 4 \int_0^\infty \Gamma_i(\tau) \cos \omega\tau \, d\tau = \frac{4q^2\overline{v_x^2}}{l} n_0 A \frac{\Theta}{1+\omega^2\Theta^2} \qquad (4\text{-}49)$$

The power density thus has a Lorentz spectrum. However, this has no significance for $\omega < 10^{12}$ rad/s as $\omega\Theta \ll 1$.

$\overline{v_x^2}$ may be expressed utilizing the equipartition theorem

$$\frac{1}{2} m^* \overline{v_x^2} = \frac{1}{2} kT \tag{4-50}$$

where $k = 1.38 \times 10^{-23}$ J/K, the Boltzmann constant. After substituting Eqs. (4-50) and (4-43), the power density of the noise current flowing through the two-pole is

$$\boxed{S_i(f) = 4kT\gamma \frac{A}{l} = 4kTG} \tag{4-51}$$

where G is the conductance of the solid shown in Fig. 4-6. This end result is the well-known Nyquist formula.

This calculation has been based on an important simplification: namely that $|v_x|$ has been regarded as constant for all charge carriers. However, actually

$$v_x = v \cos \varphi \tag{4-52}$$

where φ is the angle with the x-axis; v follows the Fermi–Dirac statistics in metals, and the Maxwell–Boltzmann statistics in semiconductors. However, for isotropic velocity distribution, Eq. (4-50) is derived again after detailed calculation. It may further be shown that for $\omega \ll 1/\Theta$, the end result is independent of the distribution of v.[8] Several other derivations of the Nyquist formula may be found in the literature.[41,126]

Expression (4-51) shows that the power density of the thermal-noise current is proportional to the conductance G and absolute temperature T of the two-pole. The power density of the noise voltage across the two-pole terminals is

$$\boxed{S_u(f) = \frac{S_i(f)}{G^2} = \frac{4kT}{G} = 4kTR} \tag{4-53}$$

$S_i(f)$, and $S_u(f)$ may be regarded as constant up to very high frequencies. However, the frequency dependence is not given precisely by Eq. (4-49) because of the approximations applied during its derivation. From the quantum mechanical model, taking into account the energy distribution and zero-point energy of the solid-state electrons,[33,142] it may be established that

$$S_i(f) = 4Ghf \left(\frac{1}{2} + \frac{1}{e^{hf/kT} - 1} \right) \tag{4-54}$$

which is identical to Eq. (4-51) for $hf \ll kT$. At room temperature, $hf = kT$ if $f \cong$

6×10^{12} Hz. Expanding Eq. (4-54) for small hf/kT values we have the approximation

$$S_i(f) = 4kTG\left[1 + \frac{1}{12}\left(\frac{hf}{kT}\right)^2 + \cdots\right] \qquad (4\text{-}55)$$

At extremely low temperatures the increase in $S_i(f)$ should be taken into account even in the microwave region. Increasing further the frequency f towards the optical regions, $S_i(f)$ is increased since the optical radiation is inherently quantized and the flow of quanta is random like the passover of charge carriers causing shot noise. This kind of noise is called *quantum noise*.

The reactances in parallel or in series with the resistive noise source have an effect on the overall power spectrum in this case too (see Sec. 3-5). However, an ideally reactive circuit element will not generate any noise as real power can neither be transmitted to nor received from its environment.

4-4 AVALANCHE NOISE

Avalanche noise may be generated between conductors in contact with each other at microscopic points, or in a p–n junction of a semiconductor with reverse bias. The breakdown mechanism of a p–n junction is fairly well known. The low-voltage breakdown below approximately 6 V is caused by the Zener (tunnel) effect while the high-voltage breakdown is the result of an avalanche-type charge carrier multiplication.[84] Avalanche noise is only generated in the latter case.[29]

There is an extremely high field strength in a p–n junction biased near breakdown. The minority carriers providing the reverse current may therefore have a high acceleration, capable of generating one or more new charge carrier pairs by ionizing on impact the neutral atoms forming the crystal lattice. Thus the plasma, characteristic of breakdown, is formed.

Figure 4-7 shows the cross section of the depletion layer of width w which is the region of high field strength in the p–n junction. The entrance of holes is at

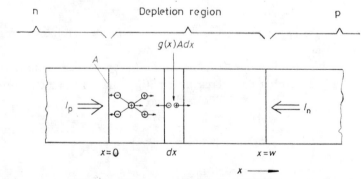

Figure 4-7 Processes in the depletion layer.

$x=0$, while that for electrons is at $x=w$. Heating or electromagnetic radiation has the effect of generating in the volume Adx further charge carrier pairs of number $g(x)Adx$ [here $g(x)$ is the number of pairs generated at point x in a unit volume and during a unit time interval]. The primary charge carriers may then generate secondary carriers during impacts, so the total current is

$$I = I_p(0)M(0)+I_n(w)M(w)+qA \int_0^w g(x)M(x)\,dx \qquad (4\text{-}56)$$

where I_p and I_n are the hole and electron currents, respectively, flowing in at the ends. $M(0)$, $M(w)$, and $M(x)$ are the multiplication factors[83] pertaining to the charge carriers entering at the ends 0 and w, or generated at point x, respectively.

Let the current increase in volume Adx be $dI(x)$, which makes the following contribution to the squared fluctuation of the total current:

$$d[\overline{i^2}(x)] = 2qM^2(x)\,dI(x)\,\Delta f \qquad (4\text{-}57)$$

in the external circuit, the squared fluctuation generated by current element $dI(x)$ will have a weighting factor $M^2(x)$ due the multiplication.

The mean square value of the total fluctuation should be calculated carefully as both $dI(x)$ and $M(x)$ are random variables. For the case where $M(x)=M$ independent of the distance and the primary (not multiplicated) current has a Poisson distribution then, according to the variance theory,[20,130] we have

$$\overline{i^2} = 2qI_0\overline{M^2}\,\Delta f \qquad (4\text{-}58)$$

where

$$I_0 = I_p(0)+I_n(w)+qA \int_0^w g(x)\,dx \qquad (4\text{-}59)$$

It has been shown by Tager[116,117] using detailed analysis of the multiplication factor M that $\overline{M^2}=\overline{M^3}$; so finally

$$S_i(f) = 2qI_0\overline{M^3} \qquad (4\text{-}60)$$

Our model may further be refined by taking into account the fact that the probability of impacts generating new charge carrier pairs is a function of the field strength, which is dependent on the distance, and is not the same for electrons and holes. If the number of new charge carriers generated in unit distance by an electron is α and by a hole is β and $\beta=k\alpha$, then according to McIntyre,[83]

$$S_i(f) = 2qI_p(0)M^3 \left[1+\frac{1-k}{k}\left(\frac{M-1}{M}\right)^2\right] \qquad (4\text{-}61)$$

if the primary current is mainly due to holes and

$$S_i(f) = 2qI_n(w)M^3\left[1-(1-k)\left(\frac{M-1}{M}\right)^2\right] \tag{4-62a}$$

or

$$S_i(f) = 2qI_n(w)M^2\left[kM+(1-k)\left(2-\frac{1}{M}\right)\right] \tag{4-62b}$$

if the primary current is mainly due to electrons. Note that the latter case is more favorable, i.e. it generates less noise than the former because for silicon $k\ll1$. This fact is actually utilized in opto-electronic diodes.[16,141]

The transition range 2–14 V between tunnel breakdown and pure avalanche breakdown has only been recently investigated.[75] It has been found that

$$S_i(f) = 2qI_0[1+f(M^*)] \tag{4-63}$$

where $M^*=(I/I_0)-1$, and $f(M^*)$ is a fairly complicated function of the averaged ionized rates. At such low voltages, the primary charge carriers are generated by the high field emission at the junction, and by these, relatively few new pairs are generated due to the short ionization distance available. The earlier theories were refined and generalized recently.[135,136]

It has not previously been taken into account that a finite time is needed for the charge carriers to pass through the depletion layer. This has the effect of making the multiplication factor time-dependent. Thus

$$M(x, \omega) = \frac{M(x)}{1+j\omega M\tau} \tag{4-64}$$

where τ is the so-called intrinsic response time,[91,98] its order of magnitude being 10^{-11} to 10^{-10} s.

However, the upper frequency limit is generally not determined by $M\tau$ but by the capacitance and series resistance of the p–n junction. As surprisingly high amplitude signals may be generated here, it is wise to apply a graphical analysis.[42] As shown by Fig. 4-8a, the junction biased up to breakdown is connected to the voltage source U_0 via current-limiting resistance R_L. Figure 4-8b shows the equivalent circuit of the junction. U_b is the breakdown voltage extrapolated for zero current, r_s is the resistance in series with the junction, and C is the total capacitance of the junction.

Figure 4-8c shows the graphic analysis of the equivalent circuit. After connecting the voltage source, capacitance C is charged to the source voltage U_0 with a time constant $R_L C$. In spite of $U_0>U_b$, breakdown will only take place if the acceleratable primary charge carrier is within the small volume of the microplasma to be formed. Because of thermal reasons, for example, this will take place after randomly changing time intervals—the earlier, on average, the higher U_0.

The avalanche-like breakdown will have the effect of inserting resistance r_s, causing a discharge of capacitance C by a time constant $(R_L \| r_s)C$ to the voltage corresponding to the stationary working point P. If the primary current, again due to the random generation process, then decreases to zero by an order of time needed to pass the depletion layer the breakdown will be terminated, as shown by one cycle illustrated in Fig. 4-8c. Figure 4-8d shows the voltage–time function across the p–n junction. Both intervals t_1 and t_2 reveal a stochastic change.

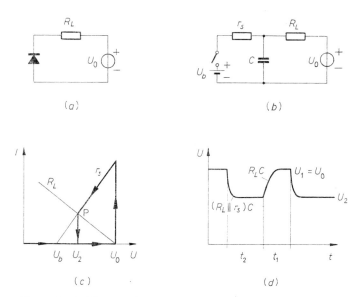

Figure 4-8 *(a)* Reverse biased p–n junction. *(b)* Equivalent circuit of the junction. *(c)* Graphical analysis of the cycle. *(d)* Voltage waveform. (*Courtesy of the American Institute of Physics and the Author.*)

The circuit shown in Fig. 4-8 and the stochastic process introduced by this circuit may also be used, under controlled conditions, as a calibrating noise source. Two possibilities for realizing this are shown in Fig. 4-9. In both cases, current generator feed is used instead of voltage source U_0 and resistance R_L. Breakdown is initiated by the field emission. In the case shown in Fig. 4-9a, the probability of field emission is given by the abruptly changing function $g_1(u)$, so the breakdowns occur by nearly equal voltages. Differentiating the stochastic triangular wave thus obtained, the power spectrum of the obtained pulses will be slightly wavy (see Fig. 4-9b). On the other hand, in the case shown in Fig. 4-9c, $g_2(u)$ is much less abrupt in making the distribution of the triangular wave cover a much wider interval. After integration, the fairly flat power spectrum (see Fig. 4-9d) is obtained.[43]

Low-voltage varactor diodes are applied as high-frequency noise sources.[40,104,110,111]

During our previous investigations, it has been implicitly assumed that the breakdown takes place simultaneously over the complete cross section of the p–n junction (uniform junction). However, this is only true if the junction is formed by a special guard ring as shown in Fig. 4-10a. In this diode, breakdown may only take place perpendicularly to the surface, in a well-defined cross section D, between layers p and n+. Reaching the state of the breakdown, the junction formed by the guard ring of low doping and the p substrate is far from breakdown.

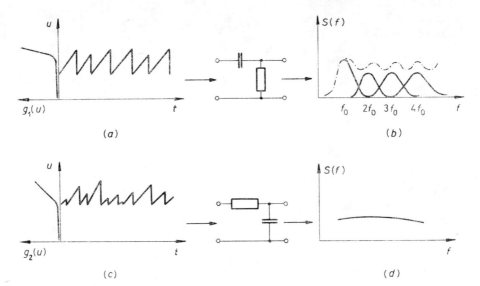

Figure 4-9 (a) Sawtooth signal with almost constant repetition rate, for a field emission probability function $g_1(u)$. (b) Power spectrum of signal a. (c) Waveform for a different type of function $g_2(u)$. (d) Power spectrum of signal c. (*Courtesy of the Institute of Electrical and Electronics Engineers.*)

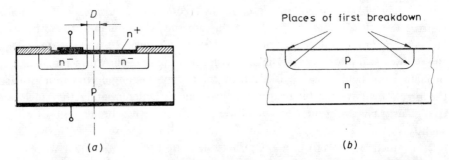

Figure 4-10 (a) Diode with guard ring for realization of the uniform junction.[43] (b) Planar diode. (*Courtesy of the Institute of Electrical and Electronics Engineers.*)

Planar diodes (Fig. 4-10*b*) may comprise two regions that may break down earlier than the planar junction. First, the breakdown voltage at curvatures is always less than at planar regions.[115, 144] Second, the perimeter where the junction reaches the surface is very delicate. If, in addition, there is some imperfect crystal structure at these sensitive points (dislocation, precipitation, or local fluctuation in the concentration of the doping), low-volume regions with sufficient field strength to generate a local avalanche—*microplasma*—may be formed.

Microplasmas have the following characteristics:

1. Generally small diameter (1–2 μm)
2. Low current intensity (of the order of 10 μA)
3. Relatively high series resistance (around 10 kΩ)
4. High noise due to stochastic in–out switching[22]
5. Occasionally weak light emission[28]

In the diodes, either a single microplasma may be generated before the complete breakdown of the junction or a multitude of microplasmas. Most of the commercially available reference diodes with more than 6-V breakdown voltage show this undesirable effect, so it is not feasible to use them with too low an operating current.

It is fairly difficult to define the parameters of the stochastic process describing the switch-on and switch-off of the microplasma. More or less general results have only been published recently.[34] It has been found that the probability of switch-on may be written in the form

$$P_{01} = [\mu i(t) + v][u(t) - U_b] \, dt \qquad (4\text{-}65)$$

where $i(t)$ is the reverse current instantaneous value, $u(t)$ is the reverse voltage instantaneous value, U_b is the breakdown voltage extrapolated for zero current, and μ and v are values characterizing the current-dependent and current-independent changeover frequencies, respectively. Let us introduce the auxiliary variable

$$z = (\mu I_0 + v) \tau_c U_0 \qquad (4\text{-}66)$$

here I_0 is the stationary reverse current, τ_c is the time constant formed by the junction capacitance and the series resistance [$\tau_c \gg \tau$, defined in Eq. (4-64)], and U_0 is the supply voltage. Utilizing this auxiliary variable, the amplitude distribution of the stochastic signal generated by a *single* microplasma is

$$F\left(\frac{u}{U_0}\right) = 1 - \frac{\Gamma[z(1 - u/U_0)]}{\Gamma(z)} \qquad (4\text{-}67)$$

Here $\Gamma(x)$ is the gamma function defined in connection with Eq. (2-118).

4-5 GENERATION–RECOMBINATION NOISE

The resistance R of a homogeneous semiconductor block of length L and cross section A is

$$R = \frac{L}{Aq(\mu_n n + \mu_p p)} = \frac{L^2}{q(\mu_n N + \mu_p P)} \tag{4-68}$$

where n and p are the densities of the electrons and holes, respectively, μ is their mobility, and N and P are the number of electrons and holes, respectively, in the *total* volume. These numbers are not constant (they may fluctuate because of the generation and recombination) so $N = N_0 + \Delta N(t)$ and $P = P_0 + \Delta P(t)$. The fluctuation is random, so the resistance will also fluctuate stochastically. Thus

$$\Delta R(t) = -\frac{L^2}{q} \frac{\mu_n \Delta N(t) + \mu_p \Delta P(t)}{\mu_n N_0 + \mu_p P_0} = -R \frac{b\Delta N + \Delta P}{b N_0 + P_0} \tag{4-69}$$

where $b = \mu_n / \mu_p$. If now a DC current I flows through the semiconductor block then the noise voltage will be

$$\Delta U(t) = -IR \frac{b\Delta N(t) + \Delta P(t)}{b N_0 + P_0} \tag{4-70}$$

For simplicity, let us investigate an intrinsic semiconductor with only pairwise recombination and generation possibilities. In this case

$$\Delta U(t) = -\frac{IR}{b N_0 + P_0} (b+1) \Delta N(t) \tag{4-71}$$

and expressing the mean square value of the fluctuation by the square of the flowing DC current, we have

$$M(\Delta I^2) = M\left(\frac{\Delta U^2}{R^2}\right) = I^2 \frac{(b+1)^2}{(b N_0 + P_0)^2} M(\Delta N^2) \tag{4-72}$$

It may be shown generally[133] that

$$M(\Delta N^2) = \frac{g(N_0)}{r'(N_0) - g'(N_0)} \tag{4-73}$$

where g is the generation rate, and g' and r' are the generation and recombination rate derivatives, respectively, in respect to N. For the intrinsic semiconductor in our example, $M(\Delta N^2) = N_0 P_0 / (N_0 + P_0)$.

Naturally, the autocorrelation function of $\Delta N(t)$ is again needed to calculate the power spectrum. The process of generation–recombination is described by a differential equation of the type given by Eq. (4-35), so it is not surprising to find that

$$\Gamma_{\Delta N}(t') = M(\Delta N^2) e^{-t'/\tau} \tag{4-74}$$

where τ is a time constant characteristic for the generation–recombination process. The power spectrum is of the Lorentz type:

$$S_{\Delta N}(\omega) = 4M(\Delta N^2) \frac{\tau}{1+\omega^2\tau^2} \qquad (4\text{-}75)$$

For a nonintrinsic-type semiconductor, the relationship $\Delta P = \Delta N$ does not hold, but these two quantities are not independent either. In this case, all components and the correlations between them have to be taken into account in the power spectrum of the current fluctuation.[132] Thus

$$S_i(f) = \left(\frac{I}{bN_0+P_0}\right)^2 [b^2 S_{nn}(f) + 2b S_{np}(f) + S_{pp}(f)] \qquad (4\text{-}76)$$

where S_{nn} and S_{pp} are the power density spectra of the electron number fluctuation and the hole number fluctuation, respectively, and S_{np} is the so-called cross power density spectrum [for its definition see Eq. (5-15)].

In semiconductors operating at not too low temperatures and having the usual level of doping, the generation–recombination noise has no significance. This is explained by the fact that in this case the conductance is determined by the density of the majority charge carriers, and this is constant being practically equal to the density of the built-in doping atoms. On the other hand, the investigation of low-concentration contaminants by evaluating the generation–recombination noise is most interesting.

During the early 1950s when the first semiconductor devices and the more-or-less pure basic materials had been prepared, the generation–recombination phenomenon and the noise due to this phenomenon had been thoroughly investigated. Interest has fallen in more recent years with the witnessing of other, stronger noise sources which suppress the former kind of noise. However, these stronger noises have been reduced by the developing technology, resulting in a renewed interest in the generation–recombination noise.[62, 131, 146] Another possibility justifying this interest is to reveal extremely low levels of contaminants or structural imperfections by careful analysis of the generation–recombination noise.

Let us take into account in the kinetics of the generation–recombination process some impurities—so-called traps. In the semiconductor under test, the density of electron traps is N_t, in addition to the completely ionized donor and acceptor impurities (see Fig. 4-11).[129] Let the density of the free electrons be n, and the density of the electrons captured by the traps be n_t. The condition of balance may be formulated in the form

Free electron + empty trap $\rightleftharpoons$ filled trap

It is assumed this time that the traps do not comprise holes or electrons from the valency band. Our results may, of course, be generalized for these cases.

Following any perturbation of the free and captured electron distribution, the system will come back to balance according to the differential equation

$$\frac{dn}{dt} = -vs(N_t - n_t)n + v_0 e^{(-W_n/kT)} n_t \tag{4-77}$$

and

$$\frac{dn_t}{dt} = -\frac{dn}{dt} \tag{4-78}$$

where v = thermal velocity of the free electrons

s = effective cross-section of the traps

$v_0 e^{-W_n/kT}$ = re-emission rate of the captured electrons.

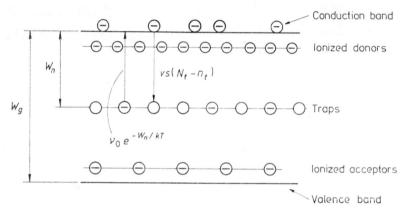

Figure 4-11 Illustration of a generation–recombination process involving traps.[129]

In the first member of Eq. (4-77), the well-known rule of the recombination is comprised: the rate is proportional to the density of unfilled traps and free electrons. The second member is introduced physically by the lattice oscillations which have the effect of freeing the captured electron from the trap. The rate v_0 has an order of 10^{12} s^{-1}.

Introducing the time constants τ_c and τ_r we have

$$\frac{1}{\tau_c} = vs(N_t - n_t) \tag{4-79a}$$

and

$$\frac{1}{\tau_r} = v_0 e^{-W_n/kT} \tag{4-79b}$$

Solving the differential equation (4-77) will give the following time function for the perturbation $\Delta n(t)$ with initial value Δn_0:

$$\Delta n(t) = \Delta n_0 e^{-t/\tau} \tag{4-80}$$

if $1/\tau = 1/\tau_c + 1/\tau_r$. Equation (4-74) may be utilized again for the characterization of the power spectrum of the process.

According to Eq. (4-79b), τ_r has a strong temperature dependence, allowing the possibility of investigating material structure by analyzing the generation–recombination noise. By measuring the time constant at several temperatures—generally attainable by cooling—and plotting this time constant as a function of $1/T$, the activation energy level W_n of the trap may be determined. On the other hand, the amplitude of the generation–recombination noise yields the density of the traps.

For more than one trap level, the time constants may not be clearly recognized in the overall noise spectrum, so the above method is only feasible with definitely separable trap levels so that the effect of the traps may be considered to be independent within the various temperature ranges. The general theory of multilevel generation–recombination fluctuation although known[133,143] is difficult to utilize in evaluating practical measurements due to mathematical complications.

The resemblance of the differential equation (4-77) with the differential equations applying to electrical networks has resulted in descriptions of the generation–recombination phenomenon by equivalent networks[25,103] and in the calculation of the generation–recombination noise by equivalent networks.[25] Figure 4-12 shows the energy diagram, the equivalent network, and the relationships between parameters applying to a semiconductor with multilevel traps. In these relations, N_i is the number of available states within the i-th level or band, f_i^0 is the equilibrium value of the probability of occupancy, and Φ_{ik}^0 [s^{-1}] is the equilibrium value of the probability of transition. Taking the conductance and valence bands as single levels, the number of total levels in the model is two more than the number of generation–recombination levels. The capacitances are proportional to the maximum number of electrons or holes within the level (band), and the conductances are proportional to the communication between these levels. The den-

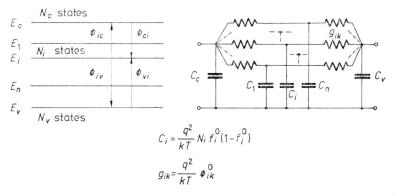

$$C_i = \frac{q^2}{kT} N_i f_i^0 (1 - f_i^0)$$

$$g_{ik} = \frac{q^2}{kT} \Phi_{ik}^0$$

Figure 4-12 Energy diagram and equivalent circuit of a multilevel generation–recombination process.

sity of charges giving rise to the conductance will only fluctuate if an electron enters or leaves the conductance band, or if a hole enters or leaves the valence band. Thus the transitions between generation–recombination levels are unimportant, so the internal nodes of the equivalent network are not connected by conductances. If it is assumed that $C_c, C_v \gg C_1, C_2, ..., C_n$, i.e. the maximum number of electrons or holes within the conductance and valence bands is high with respect to this number in the generation–recombination centers, then closed-form relationships for expressing the power spectra needed for evaluating Eq. (4-76) may be derived (with allowable approximations).[7]

To this end, the impedance matrix of the equivalent circuit shown in Fig. 4-12 has to be calculated as

$$S_{nn}(f) = N_0^2 \left(\frac{q}{kT}\right)^2 4kT \operatorname{Re} Z_{11}$$

$$S_{np}(f) = -N_0 P_0 \left(\frac{q}{kT}\right)^2 4kT \operatorname{Re} Z_{12} \qquad (4\text{-}81)$$

$$S_{pp}(f) = P_0^2 \left(\frac{q}{kT}\right)^2 4kT \operatorname{Re} Z_{22}$$

Omitting details of the calculation, we have

$$\operatorname{Re} Z_{11} = \frac{C_v/C_c}{C_c+C_v} \frac{\tau_0}{1+\omega^2\tau_0^2} \qquad (4\text{-}82)$$

$$\operatorname{Re} Z_{22} = \frac{C_c/C_v}{C_c+C_v} \frac{\tau_0}{1+\omega^2\tau_0^2} \qquad (4\text{-}83)$$

$$\operatorname{Re} Z_{12} \cong -\frac{1}{C_c+C_v} \frac{1}{(1+\omega^2\tau_0^2) \prod\limits_{1}^{n} (1+\omega^2\tau_i^2)} \sum_{i=1}^{n}\left[\frac{G_i}{G}(\tau_0+\tau_i) \prod_{k=1}^{n}{}'(1+\omega^2\tau_k^2)\right] \quad (4\text{-}84)$$

where the prime denotes the exclusion of the factor with serial number $k=i$ from the product, and $G_i = g_{ic}g_{iv}/(g_{ic}+g_{iv})$ and $G = \Sigma G_i$. Also

$$\tau_0 = \frac{C_c C_v}{C_c+C_v} \left[\sum_{i=1}^{n} G_i\right]^{-1} \qquad \tau_i = \frac{C_i}{g_{ic}+g_{iv}} \qquad (4\text{-}85)$$

A closed-form generation–recombination noise model has also been derived,[7] for another case of practical importance, the highly activated cadmium sulfide-type photoconductor.

In most kinds of solid state devices, the generation–recombination noise does not have much significance as it is usually suppressed by other kinds of noise sources. The junction field-effect transistor is usually regarded as producing the lowest noise level. However, if this device is cooled below room temperature in order to reduce the thermal noise to meet special requirements, the generation–

recombination noise may become predominant. This has been investigated in recent publications.[27,48,134] These investigations have been aimed, for instance, at proving the structural damage produced by neutron or electron irradiation,[71,140] or at demonstrating that the generation–recombination centers introduced by channel dopants are dependent on the dopant suppliers, in spite of the fact that these are so-called chemically pure elements.[47]

4-6 FLICKER NOISE

In thin metallic or semiconductor layers, in carbon resistors, carbon microphones, vacuum tubes, solid state devices, etc., another noise source in addition to those previously investigated is also present. The power spectrum of this noise source may be expressed in the general form

$$S(f) = \text{const}\, \frac{I^\beta}{f^\alpha} \qquad (4\text{-}86)$$

where I = DC current flowing through the device
f = frequency of measurement
$\beta \cong 2$
$\alpha \cong 1$

The current dependence may be easily interpreted. It will be shown that in the case of $\beta=2$, this kind of noise may be caused by the fluctuation in the conductance or resistance. On the other hand, the exponent $\alpha \cong 1$ may not presently be explained by a universal physical model, although it may be present even at a frequency of 10^{-6} Hz.[21] Furthermore, $\alpha=1$ leads also to mathematical complications: considering Problem 1-2, it is at once seen that an infinite noise power results if the lower integration limit tends to zero, and this is evidently impossible.

It is extremely difficult to give a short summary of flicker noise; this topic has been extensively dealt with in many cases. Therefore, our discussions will be restricted to investigations of homogeneous structures not comprising p–n junctions (in addition to presenting general characteristics). Also, results published since 1968 will only be presented.

In order to establish a model for conductance fluctuation, let us investigate a conductor of length l and cross section A, comprising electrons of number n_0 and mobility μ_n, per unit volume. The conductivity is then

$$\gamma = q\mu_n n_0 \qquad (4\text{-}87)$$

and

$$R = \frac{l}{\gamma A} = \frac{l}{A q\mu_n n_0} = \frac{l^2}{q\mu_n N} \qquad (4\text{-}88)$$

where $N=n_0 Al$ is the number of charge carriers within the whole volume. If N changes by dN then

$$\frac{dR}{M(R)} = -\frac{dN}{M(N)} \qquad (4\text{-}89)$$

and

$$(dR)^2 = M^2(R)\left[\frac{dN}{M(N)}\right]^2 \qquad (4\text{-}90)$$

Its expected value may be calculated from that of $(dN)^2$. Thus

$$M(dN)^2 = M[N-M(N)]^2 = CM(N) \qquad (4\text{-}91)$$

where it has been assumed that the mean square value of the fluctuation is somehow proportional to the total number of charge carriers.

From the above, we have

$$M(dR)^2 = M^2(R)\frac{C}{M(N)} = M^2(R)\frac{C}{n_0 V} \qquad (4\text{-}92)$$

that is the smaller the volume in question the higher will be the fluctuation of the resistance. This explains why this additional noise is primarily generated in thin layers and granular materials, as well as in weakly doped semiconductors (N_0 is small).

After investigating several materials and structures, Hooge[49] ventured a generalization according to which the squared relative fluctuation of a two-pole of resistance R is given by

$$\frac{\overline{\Delta R^2}}{R^2} = \frac{\alpha}{N}\frac{\Delta f}{f} \qquad (4\text{-}93)$$

where N is the total number of free charge carriers within the resistance block and α is a universal constant (2×10^{-3}). Information gathered since this publication is contradictory, and even Hooge has admitted in a publication in 1976 and comprising 87 references that "his formula may not be adapted for all inhomogeneous cases".[51]

If a current I flows through a resistance, the mean square value of the voltage fluctuation across the resistance is

$$M(u^2) = I^2 M(dR)^2 \qquad (4\text{-}94)$$

i.e. it is proportional to the current squared. However, if the voltage connected across the resistance is constant then, according to a similar calculation, the mean square fluctuation of the current flowing through the resistance is

$$M(i^2) = U^2 M(dG)^2 \qquad (4\text{-}95)$$

It is thus seen from Eq. (4-94) that the noise voltage is always generated by the current flowing through the resistance. This is why the $1/f$ noise or excess noise is frequently called current noise. However, it should be emphasized that the current is not noisy but the noise originates from the fluctuation (modulation) noise of the conductance, this being only unfolded by the current flow. (On the other hand, the generation–recombination noise is also quadratically dependent on the current, so this might also be called current noise.) The term "current noise" will therefore not be used in this book.

Of course, it should be proved that the additional noise of a resistance is actually generated by the resistance fluctuation, and that the current flow is only needed to make this fluctuation perceptible according to Eq. (4-94). It could also be possible that the current flow is already responsible for generating the fluctuation, that is the additional noise might be a result of some non-equilibrium process. However, in this case the exponent of the current in Eq. (4-94) would have to be different from two.

The validity of Eq. (4-94) has been investigated in several ways. According to one method adopted the resistance under test is considered as the series and parallel resultant of several components. This decomposition has been realized macroscopically by composing the resistance R with the aid of 1, 4, 9, 16, ... parallel and series elements.[30] The microscopic analogy of this method is the investigation of integrated circuit resistors[55] by altering the length, width, and depth. Both publications have proved the validity of Eq. (4-94).

According to another method, the exciting current I has been changed. Sutcliffe[112, 113] showed that it is relevant whether a DC or an RF current flow is applied. Jones and Francis[60] have utilized simultaneous DC and RF current flow and found a correlation factor of 0.96 ± 0.05 between the noise spectra corresponding to the respective current flows. This proves that the noise is independent of the excitation.

Other RF investigations[80, 81] have given rise to the possibility that the nonlinearity[46, 114] and/or temperature coefficient[59] may also contribute to the generation of flicker noise. On the other hand, the noise components in the frequency ranges $f \pm \Delta f$, $2f \pm \Delta f$, or $3f \pm \Delta f$ have been proved to be uncorrelated,[61] thus speaking against the nonlinearity effect. The controversy about RF investigations[13, 88] will probably last for some time.

Voss and Clarke[137, 138] have performed a basic experiment to prove that the resistance fluctuates even under conditions of thermal equilibrium, so establishing that the fluctuation is not caused by the current flow. They first investigated small volume samples, and measured $\Delta R/R$ in the usual way. Thereafter, they measured thermal noise *without current flow* which has a long time mean square value of

$$\lim_{t \to \infty} \frac{1}{t} \int_0^t U^2 \, dt' = 4kTR(f_2 - f_1) \qquad (4\text{-}96)$$

if the measurement is performed between the frequencies f_1 and f_2. If the fluctuation of R has a $1/f$ spectrum, the same spectrum should also be present at the left-hand side of the equation by choosing correct measurement intervals. The experiment proved the fluctuation of R. Other workers have obtained similar results.[10]

It is much more difficult to explain the $1/f$ frequency dependence. If the current fluctuation is explained by some kind of generation–recombination process then, according to Eqs. (4-74) and (4-75), an exponential autocorrelation function and a Lorentz power spectrum should result. Both comprise the time constant τ which is regarded as constant.

There exists a continuous distribution of the time constant which results, with good approximation, in a power spectrum of type $1/f$ in both an upper- and lower-bound frequency range. According to McWorther[85] and van der Ziel,[128] this may be conceived by the following approach.

Assume that the direct sources of the flicker noise are also generation–recombination centers which, however, are driven by slow processes caused, for example by oxide contaminations on the semiconductor surface. An exchange of charges between the semiconductor surface and the contaminant may only take place by means of the tunnel effect (see Fig. 4-13). The greater the distance y of the contaminant from the surface the more will be the average time between transitions. Thus

$$\tau = \tau_1 e^{\delta y} \tag{4-97}$$

where τ_1 is a time constant characterizing the tunnel passage, and δ is a constant of the order of $10^8 \ cm^{-1}$.

Let us divide the surface into elements of area ΔA in order to determine the distribution of time constants τ. In this case, a dominant trap within the oxide at a distance y may be assigned for each surface element. Should there be additional traps beyond these ones, these may be neglected because, according to Eq. (4-97), their time constant will increase abruptly.

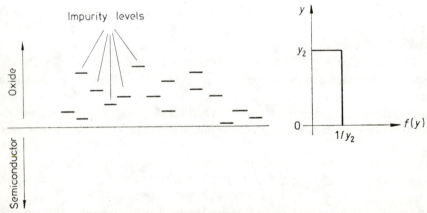

Figure 4-13 Traps in the oxide covering the surface of a semiconductor and their distribution.

Let the distribution of the traps as a function of y be uniform if $0 < y < y_2$ (see Fig. 4-13), i.e. let the density function be of the form

$$f(y) = \frac{1}{y_2} \tag{4-98}$$

From this expression, the density function of the time constant, by applying the transformation formula

$$g(\tau) = \frac{dy}{d\tau} f(y) \tag{4-99}$$

will be given, by the expression

$$g(\tau) = \begin{cases} \dfrac{1}{\delta y_2} \dfrac{1}{\tau} = \dfrac{1}{\ln \tau_2/\tau_1} \dfrac{1}{\tau} & \text{if} \quad \tau_1 < \tau < \tau_2 \\ 0 & \text{elsewhere} \end{cases} \tag{4-100}$$

Here $\tau_2 = \tau_1 e^{\delta y_2}$. It is interesting to note that by assuming the tunnel effect the reciprocal-type density function, which is quite unusual, is derived in a natural way.

The overall noise generated by the generation–recombination centers pertaining to the surface elements is determined by integrating with respect to the time constants τ. Thus

$$S(f) = 4\overline{a^2} \int_{\tau_1}^{\tau_2} \frac{\tau}{1 + \omega^2 \tau^2} \frac{d\tau/\tau}{\ln \tau_2/\tau_1} = \frac{4\overline{a^2}}{\omega \ln \tau_2/\tau_1} \int_{\omega\tau_1}^{\omega\tau_2} \frac{dx}{1 + x^2}$$

$$= \frac{4\overline{a^2}}{\omega \ln \tau_2/\tau_1} (\arctan \omega\tau_2 - \arctan \omega\tau_1) \tag{4-101}$$

In the range $1/\tau_2 \ll \omega < 1/\tau_1$,

$$S(f) \cong \frac{a^2}{f \ln \tau_2/\tau_1} \tag{4-102}$$

which is the typical $1/f$-type spectrum.

The tunnel effect may not only take place in the above model but also at the grain boundaries of resistances with granular structure, or between contact surfaces. These structures typically reveal flicker noise.

There is a high probability that the passage of electrons through the isolating layer between two metallic layers or conducting granules is made possible by the tunnel effect. This isolator is, however, seldom perfect as it may comprise traps that may be occupied by electrons (see Fig. 4-14). The height and width of the potential barrier thus formed, and so the charge quantity passed by the tunnel effect, are modulated by the occupancy of these traps. However, the occupation or depletion of the trap may only take place by the tunnel effect. Thus the distribution of time intervals between transitions here also corresponds to Eq.

(4-100).[73] According to another theorem, the modulation of the potential barrier is affected by the thermal noise of the isolator loss resistance.[66]

As the spectrum of the flicker noise is omnipresent in differing physical conditions, an attempt has been made to describe the process by purely mathematical means. Halford[44] has found a class of fluctuation phenomena meeting the following boundary conditions: only finite time constants, finite energies, and finite amplitudes are permitted. Even under these conditions, a $1/f$ spectrum of arbitrary width may be present; however, this has a lower boundary region of $\geqq 0$ exponent, and an upper boundary region of $\leqq -2$ exponent. Two other publications have utilized even more extended mathematical abstractions.[77,96]

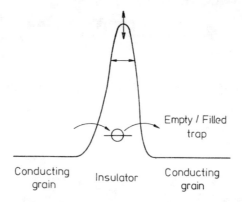

Figure 4-14 Modulation of a barrier formed in a thin insulation layer due to the varying occupancy of the trap within the insulation.

In order to discover the physical process resulting in flicker noise, numerous experiments have been carried out and several models have been created. Of more general significance are the models based on temperature fluctuation,[65,138] on mobility fluctuation,[50] and on the assumption of slow surface conditions.[85] Regarding particular structures, reference is made in the literature to metallic thin layers,[23,24,53,82,139] thick film resistances,[26,99,106,124,145] contacts,[54,121,122,123,125] and granulated structures.[1,5,6,17,45,90]

Careful investigations have revealed another peculiar characteristic of the flicker noise: its power has a substantially higher variance than the power of white noise in the same bandwidth. These investigations have been carried out on carbon resistors,[17,18,19,31,74,86,94] photoconductors,[31,52] semiconductor blocks and p–n junctions,[12,18,19,94] and operational amplifiers.[109] More or less significant deviations have been ascertained from the theoretical relative variances given by Eq. (8-6). In some papers, the discrepancy is explained by the nonstationary nature of the process;[17,31,74,118] in other sources, it is proved that the process is stationary[51,86,107,109] and the discrepancy is explained by the strong effect of small perturbations.[12, 86] It has also been investigated whether the frequency response of the measurement apparatus may be responsible for the discrepancy.[14, 39, 108] Recently the role of temperature fluctuations has also been considered.[97]

By investigating only the statistical properties of the process, the following explanations may be suggested:

1. The amplitude distribution is not gaussian.
2. The samples may not be regarded as uncorrelated.
3. The process is not stationary.

The first assumption has been disproved by many experimental results. The third assumption can neither be proved nor disproved definitely. It thus seems that a careful investigation of the second assumption may be useful.

In order to avoid analytical difficulties, a process lacking a $1/f$-type spectrum will be investigated. Fuller[37] has shown that the variance of the autocorrelation function pertaining to a process with finite length is strongly dependent on the internal characteristics of that process. For instance, consider a gaussian process of time interval T and autocorrelation function

$$\Gamma(\tau) = e^{-\lambda|\tau|} \tag{4-103}$$

The variance in this autocorrelation function is

$$\sigma_G^2(\tau) \cong \begin{cases} \dfrac{1}{2\lambda T}\left[1 - \dfrac{\tau}{2T} + (1+2\lambda\tau)e^{(-2\lambda\tau)}\right] & \text{if} \quad 0 \le \tau \le 2T \\ 0 & \text{elsewhere} \end{cases} \tag{4-104}$$

provided that $\lambda T \gg 1$.

Substituting $\tau = 0$ we get the variance of the noise power which is given here by $\sigma_G^2(0) = 1/\lambda T$. However, there may be a process that is also stationary and ergodic, but nevertheless not gaussian, in spite of the fact that it has a normal amplitude distribution. This process is illustrated by the Gauss–Poisson square wave (see Fig. 4-15). A Poisson square wave is a signal that has a level changing according to a Poisson distribution in a discrete way along the time axis; accordingly, the distribution of time intervals between changeovers is exponential. The variance of the autocorrelation function at $\tau = 0$ is

$$\sigma_{GP}^2(0) = \frac{2}{\lambda T} = 2\sigma_G^2(0) \tag{4-105}$$

The above results were first applied to high-variance noises by the author utilizing computer simulation for experimental verification.[2] It can be seen that the variance of a non-gaussian process is much higher compared with that of a pure gaussian process.

High variances have always been observed experimentally for $1/f$-type noises, so the above considerations have had to be further developed for this case too.

The analytical approximation may not be applied to this case as the autocorrelation function of the Poisson process is a simple exponential expression, while that of the flicker noise is an integral exponential or an even more complicated function, depending on the response of the band-limiting filter.[30] However, computer simulation can easily be applied. The random numbers with uniform distribution R_i are transformed by the transformation

$$t_i = \tau_1 \left(\frac{\tau_2}{\tau_1} \right)^{R_i}$$

(4-106)

into random numbers with reciprocal distribution as given by Eq. (4-100) [τ_1 and τ_2 correspond to Eq. (4-100)]. By simulating the stochastic square-wave signal according to Fig. 4-15 with such parameters and by determining the variance of the powers (mean square values), we arrived at the result[3]

$$\sigma_{GF}^2 = (5...6)\sigma_G^2$$

(4-107)

which explains the increased variance.

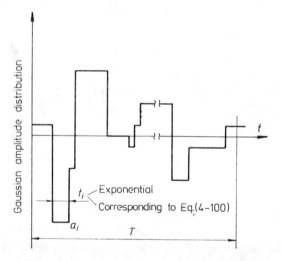

Figure 4-15 Square wave with gaussian amplitude distribution and Poisson or $1/f$-type changeover distribution.

Even more interesting is the distribution of the mean square values which, for a gaussian process and a high number of results, should also be gaussian. On the other hand, Fig. 4-16 shows a skew distribution as the result of the simulation. For comparison, the dotted line shows the distribution of the mean square values of a gaussian process having the same instantaneous values.[4] The skew distribution determined for the high-variance process should be compared with the histogram evaluating the experiments of Greenstein and Brophy.[39]

Experience shows that high-variance noise is generated if the fluctuations are the result of a series of slow effects, each effect being substantial in itself. These effects are higher than the shot-effect pulses by several orders of magnitude. Based on this observation, Fig. 4-17 shows a new physical model of a resistor generating additional noise comprising pulse noise too. Conductance modulation is given by a series of elementary, although not necessarily equal, conductances

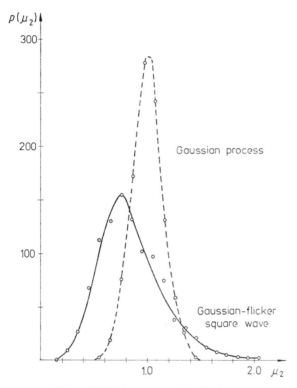

Figure 4-16 Mean square distributions.

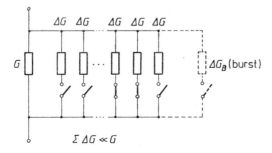

Figure 4-17 Model of resistance with additional noise, comprising a possible burst component.

that are switched in and out at random; more than one switch may change its position at a given instant.

If the ΔG values are equal then their instantaneous sum has a binomial distribution.[127] This tends to a normal distribution for a sufficiently high number of branches, thus giving a fluctuation in the overall resistance value according to the normal distribution. The time distribution of the change of states may be a distribution with a $1/f$ spectrum if the switches are driven by the above mechanism of the tunnel effect.

4-7 BURST NOISE

A pulse-type noise called burst noise or "popcorn" noise has occasionally been observed on carbon or carbon-film resistors, on p–n junctions biased into cut-off or conduction, and on discrete or integrated circuit transistors.[11, 22, 38, 72] This noise appears as a square wave with random changeovers (i.e. random telegraph signal) onto which the other noise components (thermal, shot, flicker) are superimposed (see Fig. 4-18). Occasionally, three or four levels may be distinguished instead of two levels.

Figure 4-18 Typical burst noise waveform.

It has been experimentally shown[78, 89] that the distribution functions of the time intervals t_d between the different states are

$$P_1(t_d > t) = e^{-\nu_1 t}$$
$$P_2(t_d > t) = e^{-\nu_2 t}$$

(4-108)

These functions really describe two Poisson processes. The power spectrum of the stochastic process may be determined from the Chapman–Kolmogorov equation[93]

$$s(\omega) = 4X^2 \frac{\nu_1 \nu_2}{\pi(\nu_1 + \nu_2)} \frac{1}{(\nu_1 + \nu_2)^2 + \omega^2} + 2X^2 \left(\frac{\nu_1 - \nu_2}{\nu_1 + \nu_2}\right)^2 \delta(\omega)$$

(4-109)

which is similar to the expression (3-105) describing the random symmetrical telegraph signal. The term comprising the Dirac pulse represents the DC component due to the asymmetry.

Burst noise is much less frequently observed than flicker noise, appearing generally in passive or active devices of imperfect structure. Its analysis led to several theories.[92]

Publications on burst noise in resistors are relatively rare[11, 63, 70, 86] and are rather descriptive in nature. According to our present knowledge, the overall conductance of a resistor revealing burst noise is determined by some two-state mechanism, e.g., a parallel resistor bridge that is switched on and off. The physical background of this switching process is unknown.

On the other hand, the burst noise of semiconductors has been investigated much more extensively. After proving that the time function of one avalanche-noise variant is very similar to that of the burst noise[56, 67] and thus may be separated, attention has been focused on the forward biased p–n junctions.[15, 57, 68, 69] All conclusions agree that the source of the burst noise is due to the imperfect nature of the semiconductor's crystal structure. Some workers[76, 79] claim that the presence of dislocations is sufficient to cause burst noise; in another publication,[58] a precipitation in the p–n junction is assumed to be the burst-noise source. Figure 4-19 shows the energy–band diagram of the p semiconductor–metal–n semiconductor system. The height of the potential peak at the n^+–metal junction, and thus the number of electrons capable of passing through the junction, is strongly influenced by the occupied or unoccupied state of a trap near the precipitation. As the current flowing through the junction is an exponential function of the potential peak height a single electron captured in the trap may control the movement of many other electrons.

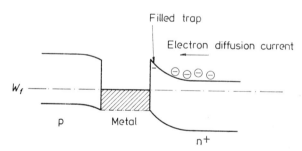

Figure 4-19 Band diagram of p–n junction comprising metal precipitation and trap. (*Courtesy of Pergamon Press.*)

The perfect periodicity of the crystal structure may be perturbed by further factors, resulting in the generation of burst noise. This kind of burst noise has been observed on samples subjected to mechanical stress,[101] dopant diffusion of high concentration,[79] thermal shock,[15] or UV irradiation.[100] The dislocation may also generate burst noise due to the mechanical stress produced by it.[32]

Burst noise was generated in a high percentage of analog integrated circuits produced in the late 1960s—a phase of rapid expansion but immature technology. This high incidence has lately been reduced to a fifth of its former value[102] using a technology that has substantially reduced the formation of metal precipitation.

PROBLEM

4-1 Assume a reverse biased semiconductor diode with a depletion layer having a width of 5 μm. The minority charge carriers pass through this depletion layer at a nearly constant velocity of 10^5 m/s. The capacity of the biased junction is 10 pF, the series diode resistance is 2 Ω. Assume an external loading circuit which is short circuited for AC. What is the frequency at which the power spectrum of the shot-noise current flowing through this circuit is decreased to half of its low-frequency value?

Answer: According to Eq. (4-20), the decrease in the power density depending on the transit time is

$$F_f^2 = \left(\frac{\sin \omega t_f/2}{\omega t_j/2}\right)^2$$

$\omega t_f/2 \cong 1.4$ pertains to $F_f^2 = 0.5$. As $t_f = 5 \times 10^{-6}$ m/10^5 m s$^{-1} = 5 \times 10^{-11}$ s, and $f = 8.9$ GHz. Neglecting the effect of the transit time, the impedance formed by the diode capacity and the series resistance also has the effect of decreasing the power density, according to Eq. (3-114). Thus

$$|H(\omega)|^2 = \frac{1}{1+(\omega r_s C)^2}$$

$f = 8$ GHz pertains to $|H(\omega)|^2 = 0.5$. The two cut-off frequencies due to two different effects are commensurable, so Eq. (4-21) could only be applicable for gross approximation. For accurate calculation, the current pulses distorted by the RC time constant should be utilized for the derivation of an expression similar to Eq. (4-20).

REFERENCES

[1] Agarwal, R. P. *et al.*: "Excess Noise in Semiconducting BaSrTiO₃", *IEEE Trans.*, vol. ED-24, no. 12, pp. 1337—1341, 1977.
[2] Ambrózy, A.: "Variance Noise", *Electr. Lett.*, vol. 13, no. 5, pp. 137—138, 1977.
[3] —: "Variance of Simulated 1/f Noise", *Electr. Lett.*, vol. 13, no. 8, pp. 224—225, 1977.
[4] —: "Variance Noise Simulations", in D. Wolf (ed.), *Noise in Physical Systems*, Springer, Heidelberg, 1978, pp. 165—168.

[5] —: "A Model for Excess Noise of Semiconducting BaSrTiO$_3$", *IEEE Trans.*, vol. ED-26, no. 9, pp. 1368—1369, 1979.

[6] —: "1/f Noise in (BaSr)TiO$_3$ (Barrier-dominated contact noise)", in E. R. Chenette (ed.) *Proceedings of the Second International Symposium on* 1/f *Noise* (Orlando-Gainesville, March 16—20, 1980), pp. 366—372.

[7] — and A. van der Ziel: "Applications of Equivalent Network Methods for Multilevel G—R Noise Spectra", *Solid State Electr.*, vol. 20, no. 5, pp. 463—467, 1977.

[8] Bakker, C. J. and G. Heller: "On the Brownian Motion in Electrical Resistances", *Physica*, vol. 6, no. 3, pp. 262—274, 1939.

[9] Beam, W. R.: *Electronics of Solids*, McGraw-Hill, New York, 1965.

[10] Beck, H. G. E. and W. P. Spruit: "1/f Noise in the Variance of Johnson Noise", *J. Appl. Phys.*, vol. 49, no. 6, pp. 3384—3385, 1978.

[11] Bell, D. A.: *Electrical Noise,* D. Van Nostrand, London, 1960.

[12] — and S. P. B. Dissanayake: "Variance Fluctuations of 1/f Noise", *Electr. Lett.*, vol. 11, no. 13, p. 274, 1975.

[13] — *et al.*: "Comments on 1/f Noise Generated by High-Frequency Currents", *Electr. Lett.*, vol. 12, no. 9, p. 235, 1976.

[14] Berz, F.: "Fluctuations in Noise Level", *Electr. Lett.*, vol. 8, no. 21, pp. 515—517, 1972.

[15] Blasquez, B. and J. Caminade: "Physical Source of Burst Noise", in D. Wolf (ed.) *Noise in Physical Systems*, Springer, Heidelberg, pp. 60—63, 1978.

[16] Brain, M. C.: "Absolute Noise Characterization of Avalanche Photodiodes", *Electr. Lett.*, vol. 14, no. 15, pp. 485—487, 1978.

[17] Brophy, J. J.: "Statistics of 1/f Noise", *Phys. Rev.*, vol. 166, no. 3, pp. 827—831, 1968.

[18] —: "Variance Fluctuations in Flicker Noise and Current Noise", *J. Appl. Phys.*, vol. 40, no. 9, pp. 3551—3553, 1969.

[19] —: "Low-Frequency Variance Noise", *J. Appl. Phys.*, vol. 41, no. 7, pp. 2913—2916, 1970.

[20] Burgess, R. E.: "Homophase and Heterophase Fluctuation in Semiconducting Crystals", *Disc. Faraday Soc.*, vol. 28, no. 3, pp. 151—158, 1959.

[21] Caloyannides, M. A.: "Microcycle Spectral Estimates of 1/f Noise in Semiconductors", *J. Appl. Phys.*, vol. 45, no. 1, pp. 307—316, 1974.

[22] Card, W. H. and P. K. Chaudari: "Characteristics of Burst Noise", *Proc. IEEE*, vol. 53, no. 6, pp. 652—653, 1965.

[23] Celasco, M. and F. Fiorillo: "Current Noise Measurements in Continuous Metal Thin Films", *Appl. Phys. Lett.*, vol. 26, no. 4, pp. 211—212, 1975.

[24] — *et al.*: "Electrical Conduction and Current Noise Mechanism in Discontinuous Metal Films", *Phys. Rev.*, vol. B17, no. 6, pp. 2553—2574, 1978.

[25] Champlin, K. S.: "Generation—Recombination Noise in Semiconductors—the Equivalent Circuit Approach", *IRE Trans.*, vol. ED-7, no. 1, pp. 29—38, 1960.

[26] Chen, T. M. and J. G. Rhee: "The Effects of Trimming on the Current Noise of Thick Film Resistors", *Solid State Techn.*, vol. 20, no. 6, pp. 49—53, 1977.

[27] Churchill, M. J. and P. O. Lauritzen: "Carrier Density Fluctuation Noise in Silicon JFETs at Low Temperatures", *Solid State Electr.*, vol. 14, no. 10, pp. 985—993, 1971.

[28] Chynoweth, A. G. and K. G. McKay: "Photon Emission from Avalanche Breakdown in Silicon", *Phys. Rev.,* vol. 102, no. 2, pp. 369—376, 1956.

[29] — and K. G. McKay: "Internal Field Emission in Silicon p-n Junctions", *Phys. Rev.*, vol. 106, no. 3, pp. 418—426, 1957.

[30] DeFelice, L. J.: "1/f Resistor Noise", *J. Appl. Phys.*, vol. 47, no. 1, pp. 350—352, 1976.

[31] Dell, R. A. *et al.*: "Experimental Study of 1/f Noise Stationarity by Digital Techniques", *J. Appl. Phys.*, vol. 44, no. 1, pp. 472—476, 1973.

[32] Doblinger, G.: Built-In Mechanical Stress as a Source of Burst Noise in Integrated Transistors", in D. Wolf (ed.): *Noise in Physical Systems*, Springer, Berlin, pp. 64—69, 1978.

[33] Ekstein, S. and N. Rostoker: "Quantum Theory of Fluctuations", *Phys. Rev.*, vol. 100, no. 4, pp. 1023—1029, 1955.

[34] Ellouze, N. *et al.:* "A Stochastic Model for Microplasma Noise", *J. Appl. Phys.*, vol. 49, no. 1, pp. 297—300, 1978.

[35] Fraser, D. B.: "Noise Spectrum of Temperature-Limited Diodes", *Wireless Eng.*, vol. 26, no. 4, pp. 129—132, 1949.

[36] Freeman, J. J.: *Principles of Noise*, John Wiley, New York, 1958.

[37] Fuller, A. T.: "Sampling Errors in the Measurement of Autocorrelation", *J. Electr. Control*, vol. 4, no. 6, pp. 551—566, 1958.

[38] Giralt, G. *et al.*: "Burst Noise of Silicon Planar Transistors", *Electr. Lett.*, vol. 2, no. 6, pp. 228—230, 1966.

[39] Greenstein, L. J. and J. J. Brophy: "Influence of Lower Cut-Off Frequency on the Measured Variance of $1/f$ Noise", *J. Appl. Phys.*, vol. 40, no. 2, pp. 682—685, 1969.

[40] Gupta, M. S.: "Noise in Avalanche Transit Time Devices", *Proc. IEEE*, vol. 59, no. 12, pp. 1674—1687, 1971.

[41] Güttler, P.: "Die Nyquistformel — Möglichkeiten zu ihrer Ableitung", *Nachrichtentechnik*, vol. 18, no. 11, pp. 425—432, 1968.

[42] Haitz, R. H.: "Model for the Electrical Behaviour of a Microplasma", *J. Appl. Phys.*, vol. 35, no. 5, pp. 1370—1376, 1964.

[43] —: "Controlled Noise Generation with Avalanche Diodes", *IEEE Trans.*, vol. ED-12, no. 4, pp. 198—207, 1965; vol. ED-13, no. 3, pp. 342—346, 1966.

[44] Halford, D.: "A General Mechanical Model for $|f|^\alpha$ Spectral Density Random Noise", *Proc. IEEE*, vol. 56, no. 3, pp. 251—258, 1968.

[45] Hanafi, H. I. and A. van der Ziel: "Flicker Noise Due to Grain Boundaries in n-Type $Hg_{1-x}Cd_xTe$", *Solid State Electr.*, vol. 21, no. 8, pp. 1019—1021, 1978.

[46] Helvoort, G. J. M. and H. G. E. Beck: "Model for the Excitation of $1/f$ Noise by High-Frequency AC Signals", *Electr. Lett.*, vol. 13, no. 18, pp. 542—543, 1977.

[47] Hiatt, C. F., Private communication, April 1976.

[48] — *et al.*: "Generation–Recombination Noise Produced in the Channel of JFETs", *IEEE Trans.*, vol. ED-22, no. 8, pp. 614—616, 1975.

[49] Hooge, F. N.: "$1/f$ Noise is No Surface Effect", *Phys. Lett.*, vol. 29A, no. 3, pp. 139—140, 1969.

[50] —: "Discussion of Recent Experiments on $1/f$ Noise", *Physica*, vol. 60, no. 1, pp. 130—144, 1972.

[51] —: "$1/f$ Noise", *Physica*, vol. 83B, no. 1, pp. 14—23, 1976.

[52] — and A. Hoppenbrouwers: "Amplitude Distribution of $1/f$ Noise", *Physica*, vol. 42, no. 2, pp. 331—339, 1969.

[53] — and T. G. M. Kleinpenning: "Comment on Current Noise Measurements in Continuous Metal Thin Films", *Appl. Phys. Lett.*, vol. 27, no. 3, p. 160, 1975.

[54] Hoppenbrouwers, A. and F. N. Hooge: "$1/f$ Noise of Spreading Resistances", *Philips Res. Rept.*, vol. 25, no. 1, pp. 69—80, 1970.

[55] Hsieh, K. C. *et al.*: "Current Noise in Surface Layer Integrated Resistors", *Solid State Electr.*, vol. 19, no. 6, pp. 451—453, 1976.

[56] Hsu, S. T.: "Bistable Noise in p-n Junctions", *Solid State Electr.*, vol. 14, no. 6, pp. 487—497, 1971.

[57] — and R. J. Whittier: "Characterization of Burst Noise in Silicon Devices", *Solid State Electr.*, vol. 12, no. 11, pp. 867—878, 1969.

[58] — *et al.*: "Physical Model for Burst Noise in Semiconductor Devices", *Solid State Electr.*, vol. 13, no. 7, pp. 1055—1071, 1970.

[59] Jones, B. K.: "$1/f$ and $1/\Delta f$ Noise Produced by a Radio Frequency Current in a Carbon Resistor", *Electr. Lett.*, vol. 12, no. 4, pp. 110—111, 1976.

[60] — and J. D. Francis: "Direct Correlation between $1/f$ and Other Noise Sources", *J. Phys. D.*, vol. 8, no. 10, pp. 1172—1176, 1975.

[61] — and J. D. Francis: "Harmonic Generation in $1/f$ Noise", *J. Phys. D.*, vol. 8, no. 16, pp. 1937—1940, 1975.

[62] Kim, S. K. *et al.*: "Noise due to Donors in *n*-Channel Silicon JFETs", *Solid State Electr.*, vol. 21, no. 9, pp. 1099—1100, 1978.

[63] Kirby, P. L.: "Current Noise in Fixed Carbon Resistors", in *Noise in Electronic Devices*, Chapman and Hall, London, pp. 78—85, 1961.

[64] Kittel, C.: *Introduction to Solid State Physics*, 2nd ed., John Wiley, New York, 1961.

[65] Kleinpenning, T. G. M.: "Temperature Fluctuation and $1/f$ Noise in Semiconductors and Metals", *Physica* B+C, vol. 84, no. 3, pp. 353—361, 1976.

[66] —: "On Low-Frequency Noise in Tunnel Diodes", *Solid State Electr.*, vol. 21, no. 7, pp. 927—931, 1978.

[67] Knott, K. F.: "Burst Noise and Microplasma Noise in Silicon Planar Transistors", *Proc. IEEE*, vol. 58, no. 9, pp. 1368—1369, 1970.

[68] —: "Instantaneous Change in Current Gain Produced by Burst Noise in a Planar Transistor", *Electr. Lett.*, vol. 13, no. 16, pp. 467—468, 1977.

[69] —: "Characteristics of Burst Noise Intermittency", *Solid State Electr.*, vol. 21, no. 8, pp. 1039—1043, 1978.

[70] Koji, T.: "Popcorn Noise and Generation–Recombination Noise Observed in Ion-Implanted Silicon Resistors", *Electr. Lett.*, vol. 11, no. 9, pp. 185—186, 1975.

[71] Krishnan, I. N. and T. M. Chen: "G–R Noise and Microscopic Defects in Irradiated Junction Field Effect Transistors", *Solid State Electr.*, vol. 20, no. 11, pp. 897—906, 1977.

[72] Leonard, P. L. and S. V. Jaskolski: "An Investigation into the Origin and Nature of Popcorn Noise", *Proc. IEEE*, vol. 57, no. 10, pp. 1786—1788, 1969.

[73] Liu, S. T. and A. van der Ziel: "Noise in Thin Film Al–Al$_2$O$_3$–Al Diodes", *Physica*, vol. 37, no. 2, pp. 241—245, 1967.

[74] López de la Fuente, J.: "Quasistationarity Noise in Resistors", *Electr. Lett.*, vol. 5, no. 12, pp. 263—265, 1969.

[75] Lukaszek, W. A. *et al.*: "Investigation of the Transition from Tunneling to Impact Ionization Multiplication in Silicon p–n Junctions", *Solid State Electr.*, vol. 19, no. 1, pp. 57—71, 1976.

[76] Luque, A. *et al.*: "Proposed Dislocation Theory of Burst Noise in Planar Transistors", *Electr. Lett.*, vol. 6, no. 6, pp. 176—178, 1970.

[77] Mandelbrot, B.: "Some Noises with $1/f$ Spectrum", *IEEE Trans.*, vol. IT-13, no. 2, pp. 289—298, 1967.

[78] Martin, J. C. and G. Blasquez: "Sur le Spectre de Bruit en Creneaux", *Solid State Electr.*, vol. 14, no. 2, pp. 89—93, 1971.

[79] — *et al.*: "L'Effet des Dislocations Cristallines sur le Bruit en Creneaux des Transistors Bipolaires au Silicium", *Solid State Electr.*, vol. 15, no. 7, pp. 739—744, 1972.

[80] May, E. J. P. and H. G. Morgan: "$1/f$ Noise in Multicarrier Systems", *Electr. Lett.*, vol. 12, no. 1, pp. 8—9, 1976.

[81] — and W. D. Sellars: "$1/f$ Noise Produced by Radio Frequency Current in Resistors", *Electr. Lett.*, vol. 11, no. 22, pp. 544—545, 1975.

[82] Mazzetti, P. and A. Stepanescu: "Mechanism of Noise Formation in Electric Conduction in Thin Metal Films", *Thin Solid Films*, vol. 13, no. 1, pp. 62—71, 1972.

[83] McIntyre, R. J.: Multiplication Noise in Uniform Avalanche Diodes", *IEEE Trans.*, vol. ED-13, no. 1, pp. 164—168, 1966.

[84] McKay, K. G.: "Avalanche Breakdown in Silicon", *Phys. Rev.*, vol. 94, no. 4, pp. 877—884, 1954.

[85] McWorther, A. L.: "1/f Noise and Germanium Surface Properties", in R. H. Kingston (ed.): *Semiconductor Surface Physics*, University of Pennsylvania Press, Philadelphia, 1957.

[86] Moore, W. J.: "Statistical Studies of 1/f Noise from Carbon Resistors", *J. Appl. Phys.*, vol. 45, no. 4, pp. 1896—1901, 1974.

[87] Moullin, E. B.: *Spontaneous Fluctuation of Voltage*, Clarendon Press, Oxford, 1938.

[88] Mueller, O.: "Comment on 1/f Noise in Resistors", *Electr. Lett.*, vol. 12, no. 2, pp. 48—49, 1976.

[89] Mulet, J. *et al.*: "Measurement of Statistical Burst Noise Characteristics", *Electr. Lett.*, vol. 6, no. 13, pp. 394—395, 1970.

[90] Mytton, R. J. and R. K. Benton: "High 1/f Noise Anomaly in Semiconducting Barium-Strontium Titanate", *Phys. Lett.*, vol. 39A, no. 4, pp. 329—330, 1972.

[91] Naqvi, I. M.: "Effects of Time Dependence of Multiplication Process on Avalanche Noise", *Solid State Electr.*, vol. 16, no. 6, pp. 19—28, 1973.

[92] Oren, R.: "Discussion of Various Views on Popcorn Noise", *IEEE Trans.*, vol. ED-18, no. 12, pp. 1194—1195, 1971.

[93] Papoulis, A.: *Probability, Random Variables and Stochastic Processes*, McGraw-Hill, New York, 1965.

[94] Purcell, W. E.: "Variance Noise Spectra of 1/f Noise", *J. Appl. Phys.*, vol. 43, no. 6, pp. 2890—2895, 1972.

[95] Rack, J. A. "Effect of Space Charge and Transit Time on the Shot Noise in Diodes", *BSTJ*, vol. 17, no. 2, pp. 592—619, 1938.

[96] Radeka, V.: "1/f Noise in Physical Measurements", *IEEE Trans.*, vol. NS-16, no. 5, pp. 17—35, 1969.

[97] Rahal, S. and A. Chovet: "Variance Noise and Temperature Fluctuations in Semiconductors", *Electr. Lett.*, vol. 15, no. 10, pp. 271—272, 1979.

[98] Ringo, J. A. and P. O. Lauritzen: "Low-Frequency White Noise in Reference Diodes", *Solid State Electr.*, vol. 15, no. 6, pp. 625—634, 1972.

[99] — *et al.*: "On the Interpretation of Noise in Thick Film Resistors", *IEEE Trans.*, vol. PHP-12, no. 4, pp. 378—380, 1976.

[100] Rodriguez, T. and A. Luque: "Behaviour of Burst Noise under UV and Visible Radiation", *Solid State Electr.*, vol. 19, no. 7, pp. 573—575, 1976.

[101] Rodriguez, T.: "Behaviour of Burst Noise under Mechanical Stress", *Electr. Lett.*, vol. 9, no. 11, pp. 248—249, 1973.

[102] Roedel, R. and C. R. Viswanathan: "Reduction of Popcorn Noise in Integrated Circuits", *IEEE Trans.*, vol. ED-22, no. 10, pp. 962—964, 1975.

[103] Sah, C. T.: "The Equivalent Circuit Model in Solid State Electronics", *Proc. IEEE*, vol. 55, no. 5, Part I, pp. 654—671, 1967; Part II, vol. 55, no. 5, pp. 672—684, 1967.

[104] Somlo, P. I.: "Zener Diode Noise Generators", *Electr. Lett.*, vol. 11, no. 14, p. 290, 1975.

[105] Spenke, E.: "Die Frequenzabhängigkeit des Schroteffektes", *Wiss. Veröff. Siemens-Werken*, vol. 16, no. 3, p. 127, 1937.

[106] Stevens, H. E. *et al.*: "High-Voltage Damage and Low-Frequency Noise in Thick-Film Resistors", *IEEE Trans.*, vol. PHP-12, no. 4, pp. 351—356, 1976.

[107] Stoisiek, M. and D. Wolf: "Recent Investigations on the Stationarity of 1/f Noise", *J. Appl. Phys.*, vol. 47, no. 1, pp. 362—364, 1976.

[108] Strasilla, U. J.: "Experimental Determination of the Variance Noise of Filtered White Noise Power Spectra", *Conference on the Physical Aspects of Noise in Semiconductor Devices*, Nordwijkkerhout, Holland, 1975, pp. 90—96.

[109] — and M. J. O. Strutt: "Narrow-Band Variance Noise", *J. Appl. Phys.*, vol. 45, no. 3, pp. 1423—1428, 1974.

[110] Susans, D. E.: "Noise Calibrator for VHF and UHF Field-Strength Measuring Receivers", *Electr. Lett.*, vol. 3, no. 8, pp. 354—355, 1967.

[111] —: "Semiconductor Diode VHF and UHF Noise Sources", *Electr. Lett.*, vol. 4, no. 4, pp. 72—73, 1968.

[112] Sutcliffe, H.: "Current Induced Resistor Noise not Attributable Entirely to Fluctuations of Conductivity", *Electr. Lett.*, vol. 7, no. 7, pp. 160—161, 1971.

[113] —: "Measurement of Fluctuations of Conductance with DC and AC Excitation", in *Conference on Le Bruit de Fond des Composants Actifs Semi-Conducteurs*, Toulouse, 1971.

[114] — and Y. Ülgen: "Spectra of AC-Induced Noise in Resistors", *Electr. Lett.*, vol. 13, no. 14, pp. 397—399, 1977.

[115] Sze, S. M. and G. Gibbons: "Effect of Junction Curvature on Breakdown Voltage in Semiconductors", *Solid State Electr.*, vol. 9, no. 9, pp. 831—845, 1966.

[116] Tager, A. S.: "Current Fluctuations in a Semiconductor under the Conditions of Impact Ionization and Avalanche Breakdown", *Fiz. Tver. Tela*, vol. 6, no. 8, pp. 2418—2427, 1964 (in Russian).

[117] — and V. M. Wald-Perlov: *Avalanche Transit Time Diodes*, Sovjetskoe Radio, Moscow, 1968 (in Russian).

[118] Tandon, J. L. and H. R. Bilger: "$1/f$ Noise as a Nonstationary Process", *J. Appl. Phys.*, vol. 47, no. 4, pp. 1697—1701, 1976.

[119] Thompson, B. J. *et al.*: "Fluctuation in Space-Charge Limited Currents at Moderately High Frequencies", *RCA Rev.*, vol. 4, no. 1, pp. 269—285, 1940.

[120] Valkó, I. P.: "Eine anschauliche Ableitung der grundlegenden Rauschformeln", *Period. Polytechn.*, vol. 15, no. 2, pp. 77—87, 1971.

[121] Vandamme, L. K. J.: "$1/f$ Noise of Point Contacts Affected by Uniform Films", *J. Appl. Phys.*, vol. 45, no. 10, pp. 4563—4565, 1974.

[122] —: "$1/f$ Noise and Constriction Resistance of Elongated Contact Spots", *Electr. Lett.*, vol. 12, no. 4, pp. 109—110, 1976.

[123] —: "On the Calculation of $1/f$ Noise of Contacts", *Appl. Phys.*, vol. 11, no. 1, pp. 89—96, 1976.

[124] —: "Criteria of Low-Noise Thick Film Resistors", *Electrocomp. Sci. Technol.*, vol. 4, July, pp. 171—177, 1977.

[125] — and R. P. Tijburg: "$1/f$ Noise Measurements for Characterizing Multispot Low Ohmic Contacts", *J. Appl. Phys.*, vol. 47, no. 5, pp. 2056—2058, 1976.

[126] van der Ziel, A.: *Noise*, Prentice Hall, Englewood Cliffs, 1954.

[127] —: *Fluctuation Phenomena in Semiconductors*, Butterworth, London, 1959.

[128] —: "Proof of Basic Semiconductor Flicker Noise Formulae", *Solid State Electr.*, vol. 17, no. 1, pp. 110—111, 1974.

[129] —: *Solid State Physical Electronics*, 3rd ed., Prentice Hall, Englewood Cliffs 1976, p. 135.

[130] —: *Noise in Measurements*, Wiley Interscience, New York, 1976.

[131] — *et al.*: "Generation–Recombination Noise at 77 °K in Silicon Bars and JFETs", *Solid State Electr.*, vol. 22, no. 2, pp. 177—179, 1979.

[132] van Vliet, K. M.: "Noise in Semiconductors and Photoconductors", *Proc. IRE*, vol. 46, no. 6, pp. 1004—1018, 1958.

[133] van Vliet, K. M. and J. R. Fassett: "Fluctuations Due to Electronic Transitions and Transport in Solids", in R. E. Burgess (ed.), *Fluctuation Phenomena in Solids*, Academic Press, New York, 1965, pp. 267—354.

[134] — and C. F. Hiatt: "Theory of Generation–Recombination Noise in the Channel of Junction Field Effect Transistors", *IEEE Trans.*, vol. ED-22, no. 8, pp. 616—617, 1975.

[135] — and L. M. Rucker: "Theory of Carrier Multiplication and Noise in Avalanche Devices, Part I", *IEEE Trans.*, vol. ED-26, no. 5, 1979, pp. 746—751.

[136] — *et al.*: "Theory of Carrier Multiplication and Noise in Avalanche Devices, Part II", *IEEE Trans.*, vol. ED-26, no. 5, 1979, pp. 752—764.

[137] Voss, R. F. and J. Clarke: "$1/f$ Noise from Systems in Thermal Equilibrium", *Phys. Rev. Lett.*, vol. 36, no. 1, pp. 42—45, 1976. .

[138] — and —: "Flicker $(1/f)$ Noise: Equilibrium Temperature and Resistance Fluctuations", *Phys. Rev.*, vol. B13, no. 2, pp. 556—573, 1976.

[139] Vossen, J. L.: "Screening of Metal Thin Film Defects by Current Noise Measurements", *Appl. Phys. Lett.*, vol. 23, no. 6, pp. 287—289, 1973.

[140] Wang, K. K. *et al.*: Neutron-Induced Noise in Junction Field Effect Transistors", *IEEE Trans.*, vol. ED-22, no. 8, pp. 591—593, 1975.

[141] Webb, P. P. *et al.*: Properties of Avalanche Diodes, *RCA Rev.*, vol. 35, no. 2, pp. 324—378, 1974.

[142] Weber, J.: "Quantum Theory of a Damped Electrical Oscillator and Noise", *Phys. Rev.*, vol. 90, no. 5, pp. 977—979, 1953; vol. 94, no. 2, pp. 211—215, 1954.

[143] Wessels, A. C. E. and K. M. van Vliet: "Some Properties of the Phenomenological Relaxation Modes of G–R Noise", *Physica,* vol. 43, no. 2, pp. 286—292, 1969.

[144] Wilson, P. R.: "Depletion Layer Calculations", *Solid State Electr.*, vol. 12, no. 1, pp. 1—12, no. 4, pp. 277—285, 1969.

[145] Yoshida, H.: "Current Crowding Effect on Current Noise on Planar Resistors", *J. Appl. Phys.*, vol. 49, no. 3, pp. 1159—1161, 1978.

[146] "International Conference on Recombination in Semiconductors", *Solid State Electr.*, vol. 21, no. 11—12, pp. 1273—1608, 1978.

FIVE

NOISE PARAMETERS
OF LINEAR NETWORKS

The noise levels due to physical processes treated in the previous Chapter are relatively low. At these levels, the electronic devices and passive network elements with good approximation may be regarded as linear. The classic mathematical methods for establishing linear network models and equivalent circuits are well known so it seems justified to apply these methods also to noisy networks.

5-1 ONE-PORTS

In the course of our investigations of the physical aspects of noise, our end results have been expressed in terms of voltage squared, current squared, or power density. The small-signal equivalent circuit of the component or device in question is made up of resistive and reactive components, so a noise equivalent circuit may be devised by including additional noise current or noise voltage generators.

Figure 5-1 shows the noise equivalent circuit of a two-pole having conductance G. If our investigations are restricted to a frequency band of width Δf and $hf \ll kT$, the mean square value of the thermal noise current is given, according to Eqs. (3-41) and (4-51), by

$$\overline{i^2} = \int_f^{f+\Delta f} S_i(f)\, df = 4kTG\Delta f \qquad (5\text{-}1)$$

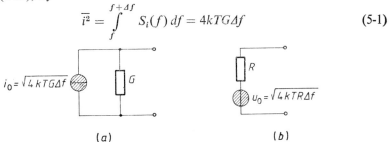

$i_0 = \sqrt{4kTG\Delta f}$ G R $u_0 = \sqrt{4kTR\Delta f}$

(a)   (b)

Figure 5-1 Real conductance noise equivalent circuits.
(a) Norton. *(b)* Thévénin.

The source current of the current generator in Fig. 5-1a is

$$i_0 = \sqrt{\overline{i^2}} = \sqrt{4kTG\Delta f} \qquad (5\text{-}2)$$

After a Norton–Thévénin transformation, we have the elements of the voltage generator equivalent circuit shown in Fig. 5-1b where

$$R = \frac{1}{G} \qquad u_0 = \sqrt{\overline{u^2}} = \sqrt{4kTR\Delta f} \qquad (5\text{-}3)$$

These equivalent circuits comprising stochastic sources are similarly applicable to those with deterministic sources. For instance, by loading the terminals of the circuit shown in Fig. 5-1b with a resistance R at temperature $T=0$ K the delivered power P is $kT\Delta f$.

As seen from Eqs. (5-2) and (5-3), the source parameters of the generators comprise the bandwidth too. This inconvenience cannot be avoided when using these equivalent circuits which are otherwise extremely useful. If $S(f)$ may be regarded as constant within a narrow frequency band, the following expressions are frequently applied:

$$i_0 = \sqrt{S_i(f)}\,\sqrt{\Delta f} \qquad [\sqrt{S_i(f)}] = \text{A/Hz}^{1/2} \qquad (5\text{-}4)$$

$$u_0 = \sqrt{S_u(f)}\,\sqrt{\Delta f} \qquad [\sqrt{S_u(f)}] = \text{V/Hz}^{1/2} \qquad (5\text{-}5)$$

The noise equivalent circuits may easily be applied to networks comprising reactive elements. The appearance of a susceptance B in parallel with conductance G will not change the source current as given by Eq. (5-1); an ideal susceptance does not generate noise. However, the terminal voltage will be frequency-dependent; according to Fig. 5-2,

$$u = \frac{i_0}{Y} = \frac{i_0}{G+j\omega C} \qquad (5\text{-}6)$$

and

$$|u(\omega)|^2 = \frac{i_0^2}{G^2+\omega^2 C^2} \qquad (5\text{-}7)$$

This latter result may be derived from Eq. (3-114) which has been obtained in another way. This means that the equivalent circuit is a true model of noisy networks even with frequency dependence.

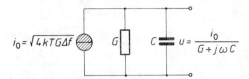

$$i_0 = \sqrt{4kTG\Delta f} \qquad G \qquad C \qquad u = \frac{i_0}{G+j\omega C}$$

Figure 5-2 Noise equivalent circuit
of admittance $Y=G+j\omega C$.

Problems arise, however, with two or more correlated noise sources within the network. In this case, the squared absolute values of the equivalent generator source parameters do not present sufficient information, and the additional information of cross correlation, which is proportional to the phases between these parameters, is also required. A simple, easily evaluated result is obtained if the overall noise voltage is calculated for a narrow frequency band.

Let $u_1(t)$ and $u_2(t)$ be two dependent stationary stochastic processes. Their sum is

$$u(t) = u_1(t) + u_2(t) \tag{5-8}$$

and the autocorrelation function of this sum, according to the definition (3-47), is

$$
\begin{aligned}
\Gamma_u(\tau) &= M[u(t)u(t+\tau)] = M\{[u_1(t)+u_2(t)][u_1(t+\tau)+u_2(t+\tau)]\} \\
&= M[u_1(t)u_1(t+\tau)] + M[u_1(t)u_2(t+\tau)+u_2(t)u_1(t+\tau)] \\
&\quad + M[u_2(t)u_2(t+\tau)] \tag{5-9} \\
&= \Gamma_{u_1}(\tau) + \Gamma_{u_1 u_2}(\tau) + \Gamma_{u_2 u_1}(\tau) + \Gamma_{u_2}(\tau)
\end{aligned}
$$

The first and last terms are the autocorrelation functions of $u_1(t)$ and $u_2(t)$, and the two middle ones are cross correlation functions according to Eq. (3-46). The latter functions are not independent of each other; this can be shown by calculating the expected value in a similar way to Eq. (3-51). Thus

$$M[u_1(t)+u_2(t+\tau)]^2 = M[u_1(t'-\tau)+u_2(t')]^2 \tag{5-10}$$

where the substitution $t'=t+\tau$ has been introduced. Performing the squaring and the calculation of the expected value, and taking into account that both processes are stationary, we have

$$
\begin{aligned}
M[u_1(t)]^2 &= M[u_1(t'-\tau)]^2 \\
M[u_2(t+\tau)]^2 &= M[u_2(t')]^2
\end{aligned} \tag{5-11}
$$

for any value of t, t', and τ. It follows from the equality of the double products that

$$\Gamma_{u_1 u_2}(\tau) = M[u_1(t)u_2(t+\tau)] = M[u_1(t'-\tau)u_2(t')] = \Gamma_{u_2 u_1}(-\tau) \tag{5-12}$$

is the internal relation, equivalent to Eq. (3-48), which was to be determined. The cross correlation function *in itself* is neither even nor odd.

The power spectrum is again calculated by the Fourier transform. By comparing Eqs. (3-62), (3-39), and (3-40), we have

$$S(f) = 2 \int_{-\infty}^{\infty} \Gamma_u(\tau) e^{-j\omega\tau} \, d\tau \tag{5-13}$$

The transforms of the autocorrelation functions $\Gamma_{u1}(\tau)$ and $\Gamma_{u2}(\tau)$ are given, as in Eq. (3-67), by

$$S_{11}(f) = 4 \int_0^\infty \Gamma_{u_1}(\tau) \cos \omega\tau \, d\tau$$

(5-14)

$$S_{22}(f) = 4 \int_0^\infty \Gamma_{u_2}(\tau) \cos \omega\tau \, d\tau$$

However, the cross correlation functions are not even functions, so

$$S_{12}(f) = 2 \int_{-\infty}^\infty \Gamma_{u_1 u_2}(\tau) e^{-j\omega\tau} \, d\tau$$

(5-15)

$$S_{21}(f) = 2 \int_{-\infty}^\infty \Gamma_{u_2 u_1}(\tau) e^{-j\omega\tau} \, d\tau$$

$$= 2 \int_{-\infty}^\infty \Gamma_{u_1 u_2}(-\tau) e^{-j\omega\tau} \, d\tau$$

$$= 2 \int_{+\infty}^{-\infty} \Gamma_{u_1 u_2}(x) e^{j\omega x}(-dx) = S_{12}(-f) = S_{12}^*(f)$$

(5-16)

where the variable substitutions $x=-\tau$ and $dx=-d\tau$ have been introduced. Finally, according to Eq. (5-9), we have

$$S(f) = S_{11}(f) + S_{12}(f) + S_{21}(f) + S_{22}(f)$$

$$= S_{11}(f) + S_{22}(f) + 2 \operatorname{Re}[S_{12}(f)]$$

(5-17)

If two stochastic processes are independent of each other then $S(f) = S_{11}(f) + S_{22}(f)$.

Let us now investigate the relationship (5-17) in an extremely narrow frequency range $(\Delta f \ll f_0)$. If all components $S_{ij}(f)$ within this range may be regarded as frequency independent then

$$\overline{u_{\Delta f}^2} = [S_{11}(f) + S_{12}(f) + S_{21}(f) + S_{22}(f)] \, \Delta f$$

(5-18)

After performing the product operation, all right-hand side terms have the dimension V^2. From these terms, the first and last terms give the "powers" of the stochastic processes $u_1(t)$ and $u_2(t)$ falling within the range Δf, which may be denoted by $|U_1|^2$ and $|U_2|^2$. (This is really the power for a load of $R=1 \, \Omega$.) S_{12} and S_{21} comprise both U_1 and U_2.

It has been shown in Sec. 3-4 that for a small relative bandwidth the stochastic signal may be approximated also by a sine (cosine) signal of frequency f_0. This is explained by following reasoning. According to Eq. (3-88), if $\Delta\omega \ll \omega_0$ and $\tau\Delta\omega \ll 2\pi$ we have

$$\Gamma_\xi(\tau) \cong A\Delta\omega \cos \omega_0 t$$

which differs from the autocorrelation function of a pure sinusoidal signal only because of the approximate nature of the equality. Let us therefore take the complex amplitude of a sinusoidal signal of frequency f_0 as $\sqrt{2}$ times U_1 and U_2; the terms of Eq. (5-18) then take the form

$$S_{11}(f)\,\Delta f = |U_1|^2 = U_1 U_1^* = |U_1|e^{j\varphi}|U_1|e^{-j\varphi}$$

$$S_{22}(f)\,\Delta f = |U_2|^2 = U_2 U_2^*$$

$$S_{12}(f)\,\Delta f = U_1 U_2^* \tag{5-19}$$

$$S_{21}(f)\,\Delta f = U_1^* U_2 = S_{12}^*(f)\,\Delta f$$

Substituting these into Eq. (5-18), we have

$$\boxed{\overline{u_{\Delta f}^2} = (U_1 + U_2)(U_1^* + U_2^*)} \tag{5-20}$$

Expression (5-20) may be generalized if the noise equivalent circuit has several voltage and/or current generators. The narrow-band overall noise voltage (or noise current) is given by the square root of the product of the linear superposition between the complex amplitudes and their conjugate values. Thus the overall noise voltage of the network shown in Fig. 5-3 is

$$u_0 = \sqrt{(U_1 + U_2 + IR)(U_1^* + U_2^* + I^* R)} \tag{5-21}$$

If the noise sources are independent then

$$u_0 = \sqrt{|U_1|^2 + |U_2|^2 + |I|^2 R^2} \tag{5-22}$$

Our relationships are valid in a chosen very narrow frequency band only. However, the calculation may be carried out for a wide-band network at separate frequencies and for linear network the results may be superimposed linearly. As the expression for calculating the overall noise voltage or current depends only on the topology of the network (the frequency may only influence the coefficients), the noise equivalent network may be utilized efficiently under rather general conditions.

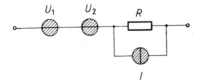

Figure 5-3 The addition
of noise powers.

5-2 TWO-PORTS

A matrix equation of the type (see Desoer and Kuh,[8] and Spence[33])

$$\begin{bmatrix} y_1 \\ y_2 \end{bmatrix} = \mathbf{M} \begin{bmatrix} x_1 \\ x_2 \end{bmatrix} + \begin{bmatrix} y^I \\ y^{II} \end{bmatrix} \tag{5-23}$$

gives the relationship between the source voltages and currents of a linear, lumped parameter two-port. Here x_1 and x_2 are the elements of the vector regarded as the independent variable, and y_1 and y_2 are elements of the vector regarded as dependent variable, $\mathbf{M}$ is the network matrix, and y^I and y^{II} are the elements characterizing the inhomogeneity. Six kinds of matrix equations may be presented depending on which voltage and/or current is taken as the independent or dependent variable. As an example, the admittance equation is

$$\begin{bmatrix} i_1 \\ i_2 \end{bmatrix} = \mathbf{Y} \begin{bmatrix} u_1 \\ u_2 \end{bmatrix} + \begin{bmatrix} i^I \\ i^{II} \end{bmatrix} \tag{5-24}$$

which may be expressed in the form

$$i_1 = Y_{11}u_1 + Y_{12}u_2 + i^I$$
$$i_2 = Y_{21}u_1 + Y_{22}u_2 + i^{II} \tag{5-25}$$

Figure 5-4 shows the definitions of these quantities. If, in addition to passive elements, the network comprises only controlled generators, then $i^I = i^{II} = 0$. Thus, the currents i^I and i^{II}, causing the inhomogeneity of the set of equations, originate from independent generators. Superscripts I and II refer to the input and output, respectively.

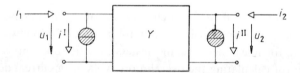

Figure 5-4 Two-port with noise sources excluded.

In all electronic devices, the DC supply power is converted into the output AC signal according to the input drive. This means that there are no independent generators in the small signal, noise-free equivalent circuit which is the linear approximation of the generally nonlinear circuit.

The noise sources may be located at different sites on the passive or active linear network. Because of the linearity requirement, their voltage or current depends only on the working point and not on the driving conditions. The overall noise voltage or current may be concentrated in an independent equivalent generator. However, there is a difference between one-ports and two-ports. A single parameter, which may be either voltage or current, is sufficient to characterize a one-port (the Norton–Thévénin conversion may be utilized for calculating one parameter from the other). However, two uncontrolled generators (e.g., i^I and i^{II}) are required to characterize a two-port. These two generators are not independent generally, and their relationship may be expressed by the complex correlation factor between them, i.e. by two data. Thus there are four data which characterize the noise conditions of linear, lumped-parameter networks.[3, 16, 18, 25, 26]

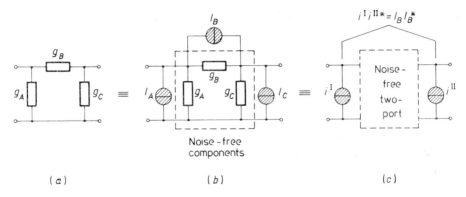

Figure 5-5 Two-port noise equivalent circuits. *(a)* Noisy two-port. *(b)* Exclusion of noise sources. *(c)* Contraction of noise sources.

To illustrate the above considerations, let us derive the noise equivalent circuit of the simple network shown in Fig. 5-5*a*. The complex amplitude of the thermal noise current of conductances g_A, g_B, g_C is given, in a very narrow frequency range Δf, by

$$I_A = \sqrt{4kTg_A \Delta f} \qquad (5\text{-}26)$$

and so on (see Fig. 5-5*b*). The source current of the input equivalent noise generator (see Fig. 5-5*c*) is determined by shorting the output and connecting an ideal current meter of zero resistance to the input. This is explained by the fact that according to Eqs. (5-25), $i_1 = i^I$ if $u_1 = u_2 = 0$. In this case, utilizing Eq. (5-20), we have

$$\overline{i_{\Delta f}^{I2}} = (I_A + I_B)(I_A^* + I_B^*) = I_A I_A^* + I_B I_B^* \qquad (5\text{-}27)$$

Similarly

$$\overline{i_{\Delta f}^{II2}} = (I_B + I_C)(I_B^* + I_C^*) = I_B I_B^* + I_C I_C^* \qquad (5\text{-}28)$$

(The second halves of the equations hold if I_A and I_B, and I_B and I_C are independent.) The correlation between the two equivalent generators is characterized by the product

$$\overline{i^I i_{\Delta f}^{II*}} = (I_A + I_B)(I_B^* + I_C^*) = I_B I_B^* \qquad (5\text{-}29)$$

The result in this case is real, $\text{Im}\,(i^{\text{I}}i^{\text{II}*})=0$. In the following consideration, the two-port block shown in Fig. 5-5c will be regarded as noise-free, as the excluded internal noise sources have been substituted by i^{I}, and i^{II}, and by their correlation.

The equivalent circuit of Fig. 5-6 may be utilized even more conveniently. The two-port, regarded as noise-free, is characterized by its chain matrix $\mathbf{L}$

$$\begin{bmatrix} u_1 \\ i_1 \end{bmatrix} = \mathbf{L} \begin{bmatrix} u_2 \\ i_2 \end{bmatrix} \tag{5-30}$$

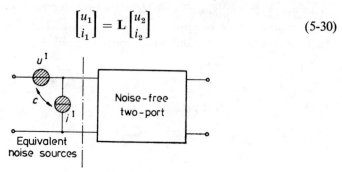

Figure 5-6 Two-generator equivalent circuit.

and the noise sources are reduced to the input. In this case, (5-30) is extended. Thus

$$\begin{bmatrix} u_1 \\ i_1 \end{bmatrix} = \mathbf{L} \begin{bmatrix} u_2 \\ i_2 \end{bmatrix} + \begin{bmatrix} u^{\text{I}} \\ i^{\text{I}} \end{bmatrix} \tag{5-31}$$

In addition to knowledge of u^{I} and i^{I}, the relationship between these quantities—their correlation factor—is also needed to be defined as

$$c = \frac{u^{\text{I}*}\,i^{\text{I}}}{\sqrt{|u^{\text{I}}|^2\,|i^{\text{I}}|^2}} \tag{5-32}$$

The two-generator noise equivalent circuits are well suited for practical applications; however, the noise bandwidth is again comprised in the source parameters as in Eqs. (5-2) to (5-5). To avoid this, the equivalent circuit based on the equivalent noise resistance and conductance, as well as the correlation admittance, has been introduced.[32]

Let us now separately investigate the noise two-port in Fig. 5-6 and denote the input quantities by subscript 1 and the output quantities by subscript 2. The quantities denoted by lower case letters are instantaneous values. The calculations may be simplified if the noise current i^{I} is resolved into two components: let i_c^{I} be the component in perfect correlation with u^{I}, and let i_n^{I} be the independent component (see Figs. 5-7a and 5-7b). Because of the perfect correlation, we have

$$i^{\text{I}} = Yu^{\text{I}} \tag{5-33}$$

where $Y_c = G_c + jB_c$ is the correlation admittance.

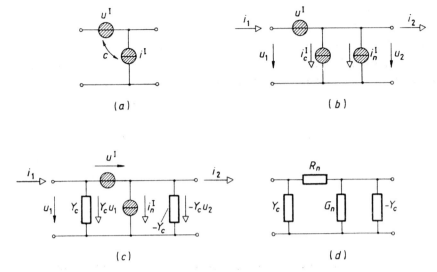

Figure 5-7 (a) Two-generator equivalent circuit. (b) Separation of the current generator into correlating and noncorrelating components. (c) Substitution of the correlating component by $Y_c u^I = Y_c(u_1 - u_2)$. (d) Resistance–conductance equivalent circuit.

In the chosen system of the measured quantities, the relationship between input and output two-port currents is

$$i_1 = i_2 + i_n^I + Y_c u^I \tag{5-34}$$

and that between the voltages is

$$u_1 = u_2 + u^I \tag{5-35}$$

Substituting this relationship back into Eq. (5-34), we have

$$i_1 = i_2 + i_n^I + Y(u_1 - u_2) \tag{5-36}$$

Equations (5-35) and (5-36) together describe the two-port shown in Fig. 5-7c where u^I and i_n^I are independent. By introducing the equivalent noise resistance

$$R = \frac{\overline{u_{\Delta f}^{I2}}}{4kT\Delta f} \tag{5-37}$$

and noise conductance

$$G_n = \frac{\overline{i_{\Delta f}^{I2}}}{4kT\Delta f} \tag{5-38}$$

we have the equivalent circuit shown in Fig. 5-7d. (The subscript Δf and the bar denoting averaging will not be used below provided that this does not introduce ambiguity.) This equivalent circuit is advantageous because R_n, G_n, G_c, and B_c do not comprise the bandwidth (although they are generally frequency-dependent). The specific circuit calculations are further simplified by contracting Y_c with the admittance of the input termination, and $(G_n - Y_c)$ with the admittance of the output termination. However, there is the disadvantage that all elements of the equivalent circuit are fictive, thus referring only to the noise and not to the useful signal. For a complicated circuit configuration, it is often difficult to decide whether certain circuit elements should be signal-related or noise-related.

In limiting cases, it sometimes occurs that only R_n of the four parameters is nonzero. For instance, this is the situation in the low-frequency investigation of grid valves, and in this case, the equivalent circuit of Fig. 5-7d is advantageous.

5-3 NOISE FIGURE

Information transmission systems comprise a signal source (transmitter), a two-port or several two-ports realizing the transmission channel, and a signal utilizer (receiver). In the following example the signal source will be regarded as a one-port for simplicity. The acousto-electrical, the opto-electronic and similar converters are really two-ports[29, 35-37] but they are still characterized by their equivalent noise resistance. If the two-port has a power gain of A_p, then A_p times the signal source noise power will appear at the two-port output; however, the noise power generated within the two-port will always be added to this output.[19] The noise figure is used to characterize the ratio of these noise powers of different origins

$$F = \frac{P_{no}}{A_p P_{ni}} = 1 + \frac{P'_{n0}}{A_p P_{ni}} \qquad (5\text{-}39)$$

where P_{no} is the total output noise power, and P'_{no} is the fraction of this power generated within the two-port. This relation clearly shows that the noise figure may only be defined for a signal source having finite resistance, i.e. $P_{ni} \neq 0$. It is seen that $F_{min} = 1$. Generally, the logarithmic presentation

$$F_{\log} = 10 \log F \quad (\text{dB}) \qquad (5\text{-}40)$$

is used. If the second term in Eq. (5-39) is much less than one, the dB value is not significant. In this case, the concept of noise temperature is introduced. The overall noise figure reduced to the input is

$$\frac{P_{no}}{A_p} = P_{ni} + \frac{P'_{no}}{A_p} \qquad (5\text{-}41)$$

In a matched case, we have

$$P_{ni} = kT\Delta f \qquad (5\text{-}42)$$

As the noise power reduced to the input is higher than P_{ni}, Eq. (5-41) may be expressed as

$$\frac{P_{no}}{A_p} = k\Delta f(T + \Delta T) \qquad (5\text{-}43)$$

where ΔT is the equivalent noise temperature of the two-port. However, this parameter should be applied with caution if different parts of the circuit have different temperatures.[7]

The noise figure may easily be determined with the aid of the equivalent circuits introduced in Sec. 5-2. According to Fig. 5-6, the noisy two-port may be regarded as the cascade connection of a two-port comprising two partially correlated noise generators and a noise-free two-port, thus the latter part may be disregarded for the noise figure calculation. Figure 5-8 shows the equivalent two-port comprising the generators with internal impedance Z_g, supplying a noise voltage of instantaneous value u^l and a noise current of instantaneous value i^l, and having a power gain $A_p = 1$; $(u^l = i^l = 0$ means a direct through connection).

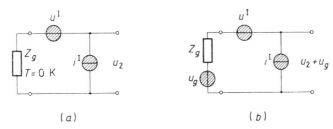

(a)

(b)

Figure 5-8 Equivalent circuits for noise figure calculation
$$(u_2 + u_g)(u_2^* + u_g^*) = |u_2|^2 + |u_g|^2.$$

In order to calculate the noise figure, let us first determine the squared absolute value of the open circuit output voltage u_2 when Z_g is noiseless (see Fig. 5-8a). As in Eq. (5-20),[27] we have

$$|u_2|^2 = (u^l + Z_g i^l)[u^{l*} + (Z_g i^l)^*] = |u^l|^2 + |Z_g i^l|^2 + 2 \operatorname{Re}(u^{l*} Z_g i^l) \qquad (5\text{-}44)$$

Let the generator impedance be
$$Z_g = R_g + jX_g \qquad (5\text{-}45)$$

and the correlation factor be
$$c = a + jb = \frac{u^{l*} i^l}{\sqrt{|u^l|^2 |i^l|^2}}$$

[see Eq. (5-32)]. Equation (5-44) may then be expressed as

$$|u_2|^2 = |u^l|^2 + |Z_g i^l|^2 + 2\sqrt{|u^l|^2 |i^l|^2}(a R_g - b X_g) \qquad (5\text{-}46)$$

Actually Z_g can also generate a noise voltage of mean square value $u_g^2 = 4kTR_g\Delta f$. According to Fig. 5-8b, this is added to $|u_2|^2$, so the noise figure becomes

$$F = \frac{|u_g|^2 + |u_2|^2}{|u_g|^2} = 1 + \frac{|u_2|^2}{|u_g|^2} \qquad (5\text{-}47)$$

For any (noise-free) load resistance connected to the output terminals of the two-port shown in Fig. 5-8b, the ratio of $|u_2|^2$ to $|u_g|^2$ remains unchanged. [u_g and u_2 are generated at independent sources and their correlation is zero, so in the numerator of Eq. (5-47) there appears the sum of their squared absolute values.] By substituting Eqs. (5-3) and (5-46), we have

$$F = 1 + \frac{|u^{\mathrm{I}}|^2 + R_g^2 |i^{\mathrm{I}}|^2 + X_g^2 |i^{\mathrm{I}}|^2 + 2\sqrt{|u^{\mathrm{I}}|^2 |i^{\mathrm{I}}|^2} (aR_g - bX_g)}{4kT R_g \, \Delta f} \tag{5-48}$$

This means that a knowledge of the four noise parameters (u^{I}, i^{I}, a, b) of the two-port and the input termination allows the noise figure to be calculated. The source parameters u^{I} and i^{I} have to be expressed for the bandwidth Δf in the denominator. Thus the application of this formula is somewhat inconvenient, as is the case of Eqs. (5-2) and (5-3).

A minimum noise figure may be attained by changing R_g. Differentiating Eq. (5-48) with respect to R_g we obtain

$$R_{g\,\mathrm{opt}}^2 = \frac{|u^{\mathrm{I}}|^2 + X_g^2 |i^{\mathrm{I}}|^2 - 2bX_g \sqrt{|u^{\mathrm{I}}|^2 |i^{\mathrm{I}}|^2}}{|i^{\mathrm{I}}|^2} \tag{5-49}$$

For $X_g = 0$

$$R_{g\,\mathrm{opt}} = \left| \frac{u^{\mathrm{I}}}{i^{\mathrm{I}}} \right| \tag{5-50}$$

independently from the correlation factor c. In this case

$$F_{\min} = 1 + \frac{|u^{\mathrm{I}}|^2 (1+a)}{2kTR_g \, \Delta f} \tag{5-51}$$

If, on the other hand, $X_g \neq 0$, but $b = 0$

$$R_{g\,\mathrm{opt}}^2 = \left| \frac{u^{\mathrm{I}}}{i^{\mathrm{I}}} \right|^2 + X_g^2 \tag{5-52}$$

and

$$F_{\min} = 1 + \frac{|u^{\mathrm{I}}|^2 + |i^{\mathrm{I}}|^2 \left(X_g^2 + aR_g \sqrt{R_g^2 - X_g^2} \right)}{2kTR_g \, \Delta f} \tag{5-53}$$

Expressions (5-50) to (5-53) are primarily intended for low-frequency noise calculations.

Figure 5-7 shows the equivalence of the two-generator equivalent circuit and the noise resistance–conductance equivalent circuit. This means that the noise figure may also be expressed by the parameters of the latter circuit. The short-circuit output current of the two-port shown in Fig. 5-9, for $i_g = 0$, is given by

$$i_2 = i_n^{\mathrm{I}} + u^{\mathrm{I}}(Y_g + Y_c) \tag{5-54}$$

and

$$|i_2|^2 = |i_n^{\mathrm{I}}|^2 + |u^{\mathrm{I}}|^2 |Y_g + Y_c|^2 \tag{5-55}$$

$$Y_g = G_g + jB_g \qquad u^{\mathrm{I}} = \sqrt{4kTR_n \Delta f}$$

$$Y_c = G_c + jB_c \qquad i_n^{\mathrm{I}} = \sqrt{4kTG_n \Delta f}$$

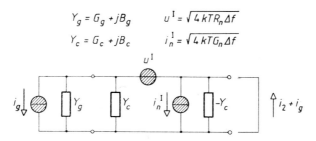

Figure 5-9 Noise figure calculation from the equivalent circuit shown in Fig. 5-7c.

as i_n^{I} and u^{I} are independent according to Sec. 5-2. On the other hand, if only the noise of the signal source is taken into account, then $i_n^{\mathrm{I}} = u^{\mathrm{I}} = 0$, the output current is i_g, and the noise figure, as in Eq. (5-37), is

$$F = 1 + \frac{|i_2|^2}{|i_g|^2} = 1 + \frac{|i_n^{\mathrm{I}}|^2}{|i_g|^2} + \frac{|u^{\mathrm{I}}|^2}{|i_g|^2}|Y_g + Y_c|^2 = 1 + \frac{G_n}{G_g} + \frac{R_n}{G_g}|Y_g + Y_c|^2 \quad (5\text{-}56)$$

where Eqs. (5-37) and (5-38) have been used. Resolving the admittances into their real and imaginary parts, we have

$$\boxed{F = 1 + \frac{G_n}{G_g} + \frac{R_n}{G_g}[(G_g + G_c)^2 + (B_g + B_c)^2]} \qquad (5\text{-}57)$$

If $B_g = B_c = 0$ or $B_g = -B_c$, then

$$F = 1 + \frac{G_n}{G_g} + \frac{R_n}{G_g}(G_g + G_c)^2 \qquad (5\text{-}58)$$

The minimum value of this noise figure is obtained by choosing

$$G_{g\,\mathrm{opt}}^2 = \frac{G_n}{R_n} + G_c^2 \qquad (5\text{-}59)$$

The noise figure then becomes

$$F_{\min} = 1 + 2R_n(G_{g\,\mathrm{opt}} + G_c) \qquad (5\text{-}60)$$

By choosing $B_{g\,\mathrm{opt}} = -B_c$, the subtraction of Eq. (5-60) from (5-57) yields an easily applicable expression:

$$F - F_{\min} = \frac{G_n}{G_g} + \frac{R_n}{G_g}[(G_g + G_c)^2 + (B_g - B_{g\,\mathrm{opt}})^2] - 2R_n(G_{g\,\mathrm{opt}} + G_c)$$

$$= \frac{R_n}{G_g}[(G_g - G_{g\,\mathrm{opt}})^2 + (B_g - B_{g\,\mathrm{opt}})^2] \qquad (5\text{-}61)$$

Equation (5-61) is an expression of the noise matching. The noise figure will increase if G_g or B_g differs from the optimum value. The generator admittance pertaining to the minimum noise figure does not necessarily provide the best power match. In high-frequency amplifiers, B_c is often significant—primarily because of capacitive feedthrough—so the conditions for optimum power transfer and minimum noise figure are widely different.

The realization of optimum noise match requires varied consideration.[10, 30] Obvious solutions, like the application of a transformer for matching purposes, do not always turn out to be the best compromise. For instance, if the source impedance is low, the match may be accomplished by the parallel connection of several similar active devices to form the amplifier input stage.[23] Separate investigations are required for matching signal sources which produce nonwhite (e.g. flicker) noise.[2]

5-4 COMPLEX NETWORKS

From the noise aspect, the cascaded and feedback amplifiers are the most important networks. In the case of the cascade connection, the overall noise figure is, in most cases, primarily determined by the noise figure of the first stage. The feedback in itself does not alter the noise relationship. However, modification of the impedances may bring the conditions of optimum noise match and optimum power match closer to each other. The internal noise of the feedback network will increase the overall noise figure.

Figure 5-10 shows cascaded two-ports. A match is assumed at all connection points. If the signal source supplies a noise power of $P_{ni}=kT\Delta f$ and the transferred noise powers at the two-port connection points are $P_{no1}, P_{no2}, \ldots,$ then

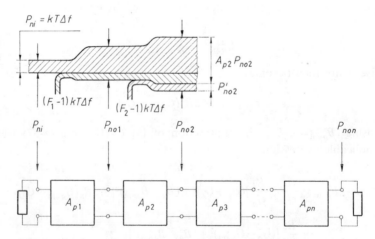

Figure 5-10 Calculation of overall noise figure of cascaded two-ports.

the following relationships are valid:

$$P_{no2} = A_{p2}P_{no1} + P'_{no2}$$

$$P_{non} = A_{pn}A_{p(n-1)} \cdots A_{p2}P_{no1} + A_{pn}A_{p(n-1)} \cdots A_{p3}P'_{no2}$$

$$+ \ldots + A_{pn}P'_{no(n-1)} + P'_{non} \tag{5-62}$$

The output of the second two-port appears as A_{p2} times the power originating from the first two-port and the noise power originating in the second two-port, and so on. As the two-port input has a matched termination, the latter can be expressed, according to Eq. (5-39), in the form

$$P'_{no2} = (F_2 - 1)A_{p2}kT\Delta f$$

$$P'_{non} = (F_n - 1)A_{pn}kT\Delta f \tag{5-63}$$

In a similar way

$$P_{no1} = F_1 A_{p1} kT\Delta f \tag{5-64}$$

The overall noise figure of the cascade connected two-ports is calculated by substituting Eqs. (5-63) and (5-64) into (5-62) and dividing by the product of the input noise power and power gain. Thus

$$F = \frac{P_{non}}{P_{ni}\prod_1^n A_{pi}} = F_1 + \frac{F_2 - 1}{A_{p1}} + \frac{F_3 - 1}{A_{p1}A_{p2}} + \ldots + \frac{F_n - 1}{\prod_1^{n-1} A_{pi}} \tag{5-65}$$

The higher the first power gain the less the effect of the later stages of the chain on the overall noise figure.[13] It follows from this statement that attenuating networks should be avoided in the first stages.

It is also seen from Eq. (5-65) that it is not sufficient to characterize a two-port with a single parameter for noise purposes. If, for instance, there is a simple through connection or an ideal transformer at the first place in the chain, $F_1 = 1$ but A_{p1} is also one, so the overall noise figure will primarily be determined by F_2 . The practical question is formulated by asking that if two different stages are given, which of the two stages is to be placed at the beginning of the chain?

Let the overall noise figure of the cascaded two-ports 1 and 2 be F_{12}; let F_{21} denote the reverse case. Then

$$F_{12} = F_1 + \frac{F_2 - 1}{A_{p1}} \tag{5-66}$$

and

$$F_{21} = F_2 + \frac{F_1 - 1}{A_{p2}} \tag{5-67}$$

Let us further assume that $F_{12} < F_{21}$. By adding -1 to both sides of Eqs. (5-66) and (5-67) and rearranging, we have

$$(F_1 - 1) - \frac{F_1 - 1}{A_{p2}} < (F_2 - 1) - \frac{F_2 - 1}{A_{p1}} \tag{5-68}$$

After further rearrangement, our condition takes the form

$$M_1 = \frac{F_1-1}{1-1/A_{p1}} < \frac{F_2-1}{1-1/A_{p2}} = M_2 \qquad (5\text{-}69)$$

The fractions in expression (5-69) are called noise measures, denoted by M.[14,17,30] The more F_1 approximates unity and the more the power gain, the less is the noise measure. For $F_1 = 1 + \Delta F_1$ and $A_{p1} \gg 1$, the noise measure is approximately given by

$$M_1 \approx \Delta F_1 \left(1 + \frac{1}{A_{p1}}\right) \qquad (5\text{-}70)$$

which is directly suitable for the comparison of amplifiers.

A feedback amplifier is derived by connecting in series and/or in parallel with the amplifier input and output ports a feedback two-port. Figure 5-11 demonstrates the possible configurations, showing for each combination the parameter system resulting in the simplest calculation.[8,20,33]

According to Fig. 5-6, the amplifier noise sources may be taken into account by applying a two-generator equivalent circuit. These generators are within the feedback loop, so their effect at the output changes by the same amount as the amplification, in a similar way to the distortion or hum effect; the effect decreases

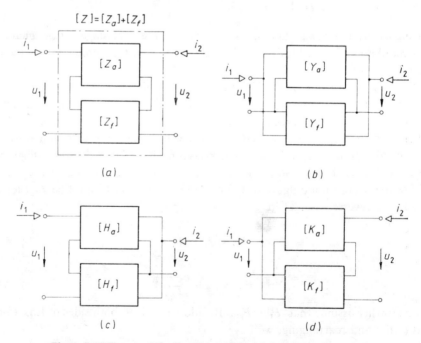

Figure 5-11 Types of feedback networks. *(a)* Series–series type. *(b)* Parallel–parallel type. *(c)* Series–parallel type (*H* hybrid matrix). *(d)* Parallel–series type (*K* inverse hybrid matrix).

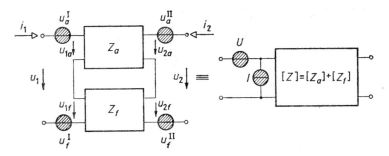

Figure 5-12 Calculation of noise parameters of series–series
feedback-type amplifier.

in the case of negative feedback.[4] This means that as a first approximation the
noise figure is not affected by feedback.

For a more thorough analysis, the noise sources within the feedback network
must also to be taken into account.[11, 12, 20, 22] As an example, let us consider the
series–series feedback (see Fig. 5-12). In this case, the impedance matrices of
the two networks form a suitable starting point. As in Eq. (5-23), we have

$$\begin{bmatrix} u_{1a} \\ u_{2a} \end{bmatrix} = Z_a \begin{bmatrix} i_1 \\ i_2 \end{bmatrix} + \begin{bmatrix} u_a^I \\ u_a^{II} \end{bmatrix} \tag{5-71}$$

and

$$\begin{bmatrix} u_{1f} \\ u_{2f} \end{bmatrix} = Z_f \begin{bmatrix} i_1 \\ i_2 \end{bmatrix} + \begin{bmatrix} u_f^I \\ u_f^{II} \end{bmatrix} \tag{5-72}$$

where the subscripts a and f refer to the amplifier and feedback networks, respectively. The impedance matrix of the overall two-port is

$$\mathbf{Z} = \mathbf{Z}_a + \mathbf{Z}_f \tag{5-73a}$$

and the input and output equivalent noise voltage generators are

$$u^I = u_a^I + u_f^I \quad \text{and} \quad u^{II} = u_a^{II} + u_f^{II} \tag{5-73b}$$

Usually, the noise equivalent circuit shown in Fig. 5-6 is known for the amplifier
and feedback networks, with parameters now denoted by U_a, I_a, U_f, and I_f, so
for calculating the quantities on the right-hand sides of Eqs. (5-73b), the following
transformation has to be used:

$$\begin{bmatrix} u_{a,f}^I \\ u_{a,f}^{II} \end{bmatrix} = \begin{bmatrix} 1 & -Z_{11a,f} \\ 0 & -Z_{21a,f} \end{bmatrix} \begin{bmatrix} U_{a,f} \\ I_{a,f} \end{bmatrix} \tag{5-74}$$

On the other hand, u^I and u^{II} calculated by Eqs. (5-73b) may be transformed
back using the transformation

$$\begin{bmatrix} U \\ I \end{bmatrix} = \begin{bmatrix} 1 & -Z_{11}/Z_{21} \\ 0 & -1/Z_{21} \end{bmatrix} \begin{bmatrix} u^I \\ u^{II} \end{bmatrix} \tag{5-75}$$

By combining Eqs. (5-73) to (5-75), we have

$$U = U_a + I_a[Z_{11}(Z_{21a}/Z_{21}) - Z_{11a}] + u_f^{\mathrm{I}} - u_f^{\mathrm{II}}(Z_{11}/Z_{21}) \qquad (5\text{-}76)$$

$$I = I_a - u_f^{\mathrm{II}}(1/Z_{21}) \qquad (5\text{-}77)$$

In Eq. (5-76), the multiplier of I_a may be rewritten as

$$Z_{11}\frac{Z_{21a}}{Z_{21}} - Z_{11a} = \frac{Z_{11a}Z_{11f}}{Z_{21a} + Z_{21f}}\left[\frac{Z_{21a}}{Z_{11a}} - \frac{Z_{21f}}{Z_{11f}}\right] \qquad (5\text{-}78)$$

Z_{21}/Z_{11} is the voltage gain of the two-port characterized by the impedance parameters, so from Eq. (5-78), we have

$$Z_{11f}\frac{A_a - A_f}{A_a + (Z_{11f}/Z_{11a})A_f} \cong Z_{11f} \qquad (5\text{-}79)$$

as $A_a \gg A_f$, and generally $Z_{11f} < Z_{11a}$. In the last term of Eq. (5-76), Z_{11}/Z_{21} is the reciprocal value of the feedback amplifier voltage gain which is less than one; thus this term may also be neglected. The second term of Eq. (5-77) is also much less than the first term. Accordingly, the equivalent noise sources of the feedback amplifier are

$$U \cong U_a + I_a Z_{11f} + u_f^{\mathrm{I}} \qquad (5\text{-}80)$$

$$I \cong I_a \qquad (5\text{-}81)$$

The same result is obtained after the rearrangement shown in Fig. 5-13.

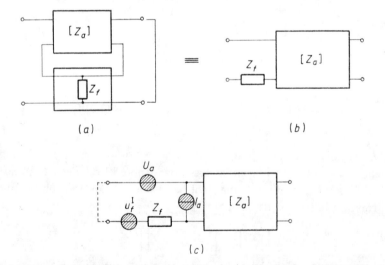

(a)

(b)

(c)

Figure 5-13 General procedure for reducing to the input port the most important noise parameter of a feedback two-port. (a) A short circuit is applied in place of the load, and the amplifier output is disconnected. (b) After redrawing. (c) All noise sources reduced to the input. (The procedure may be applied to all configurations shown in Fig. 5-11.)

As regards the noise relations, Z_f is series connected with the amplifier input, so the thermal noise of $\mathrm{Re}\,(Z_f)$ and the additional noise voltage drop introduced by I_a have to be taken into account. Naturally, the latter correlates completely with I_a. The correlation factor c between U and I is

$$c = \frac{(U_a+I_a\overline{Z_{11f}+u_f^1})^*\,I_a}{\sqrt{|U|^2|I|^2}} = \frac{U_a^*\,I_a+|I_a|^2Z_{11f}}{\sqrt{|U|^2|I^2|}} = \frac{c_a\sqrt{|U_a|^2|I_a|^2}+|I_a|^2Z_{11f}}{\sqrt{|U|^2|I|^2}} \quad (5\text{-}82)$$

where c_a is the correlation factor between the noise generators of the non-feedback amplifier. Relationships similar to Eqs. (5-80) to (5-82) may be derived for several configurations of Fig. 5-12. Elements of the feedback two-port of importance for noise considerations may be reduced to the input using the procedure shown in Fig. 5-13. The network thus obtained contains a series impedance Z_s and/or a parallel impedance Z_p (see Fig. 5-14a). The equivalent noise generators of the non-feedback system (see Fig. 5-14b), from knowledge of the chain matrix of this network and of the equivalent noise generators of the non-feedback amplifier, are

$$U = a_{11}[U_a+u_s^1]+a_{12}I_a \quad (5\text{-}83)$$

$$I = a_{22}[I_a+i_p^1]+a_{21}U_a \quad (5\text{-}84)$$

where

$$u_s^1 = \sqrt{4kT\varDelta f\,\mathrm{Re}\,(Z_s)} \quad (5\text{-}85)$$

$$i_p^1 = \sqrt{4kT\varDelta f\,\mathrm{Re}\,(1/Z_p)} \quad (5\text{-}86)$$

The overall correlation factor may be determined as in Eq. (5-82).

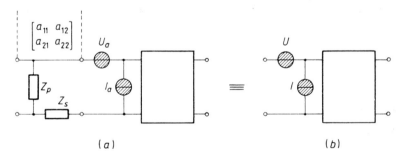

(a)   (b)

Figure 5-14 (a) Elements of the feedback network reduced to the input.
(b) Resultant noise generators.

Application of feedback in an input stage is advantageous if a change in the input impedance is advisable as, for instance, in the case of noise figure optimization,[10] and the increase in noise figure due to the correctly designed feedback network is negligible. Special attention should be paid to the operational amplifier which is practically always used with feedback. In order to decrease the offset voltage, the overall value of the resistances connected to the inputs must to be equal (i.e. symmetrical).

Unfortunately, the resistances R_1 and R_2 inserted in order to decrease the offset voltage substantially increase the overall noise voltage, as shown by the following expressions:[24]

1. For a noninverting amplifier

$$u_n^2 = 4kT\Delta f\left[R_1 + R_2\left(1 + \frac{A-1}{A^2}\right)\right] + u^{l2} + i^{l2}(R_g + R_1 + R_2)^2 \qquad (5\text{-}87)$$

2. For an inverting amplifier

$$u_n^2 = 4kT\Delta f\left[R_1\left(1 + \frac{1}{|A|}\right) + R_2\left(1 + \frac{1}{|A|}\right)^2\right] + u^{l2}\left(1 + \frac{1}{|A|}\right)^2 + i^{l2}(R_g + R_1 + R_2)^2 \qquad (5\text{-}88)$$

where u_n is the overall noise voltage reduced to the amplifier input, A is the actual amplification according to Fig. 5-15, and u^l and i^l correspond to the previous interpretation; these comprise the nonwhite (e.g., flicker) noise of the amplifier too.

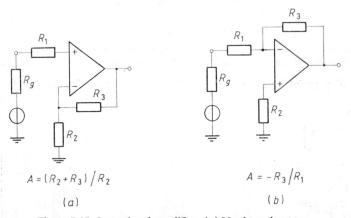

$$A = (R_2 + R_3)/R_2 \qquad\qquad A = -R_3/R_1$$

$$(a) \qquad\qquad\qquad (b)$$

Figure 5-15 Operational amplifier. *(a)* Noninverting type. *(b)* Inverting type.

Calculation of the overall output noise voltage for a complicated linear network generally requires a computer. Any circuit analysis program is theoretically suitable because the noise sources may be taken into account as independent current generators connected in parallel with the network branches or as series voltage generator; their effect may be added at the output. To increase the speed of this procedure, several methods have been devised. The first was the adjunct network method based on interreciprocity which was derived by a suitable transformation of the matrix describing the original network.[9,31] It was later shown that the overall noise voltage may be calculated even more simply by utilizing the network element sensitivities.[5,15,21] Sensitivity calculations are included anyway in a modern network analysis program, so the noise analysis is virtually performed without any additional cost. Naturally, a condition for this is the inclusion of the noise parameters in the network elements data; in many cases, however, these values may only be determined empirically.[6,28]

5-5 NOISE BANDWIDTH

The equivalent circuits given in previous Sections of this Chapter apply to an infinitely small range around a selected frequency. As the networks are linear, several results pertaining to small adjoining frequency ranges may be added. Let us assume that the bandwidth of the overall frequency range is f_2-f_1, the transfer function of the two-port is $H(f)$ and the one-sided power density of the noise source connected to the input is $S(f)$. Then according to Eq. (3-114), the squared output noise voltage is

$$\overline{U_{\text{out}}^2} = \int_{f_1}^{f_2} S(f)|H(f)|^2\, df \tag{5-89}$$

Equation (5-89) may be generalized for a network comprising several noise sources. If, for instance, a two-port with a transfer function $H(f)$ is connected to the output port of the network shown in Fig. 5-8a, the squared output noise voltage of this two-port is

$$\overline{U_{\text{out}}^2} = \int_{f_1}^{f_2} \{S_u(f)+|Z_g(f)|^2 S_i(f)+2\,\text{Re}\,[Z_g(f)S_{ui}(f)]\}|H(f)|^2\, df \tag{5-90}$$

as in Eq. (5-44). As correlation may exist between u and i, $S_{ui}(f)$, which is the cross power density according to Eq. (5-15), appears in Eq. (5-90). It may be seen from Eqs. (5-89) and (5-90) that $\overline{U_{\text{out}}^2}$ depends not only on the frequency response of the transfer function but also on the spectrum of the driving signal.

Let $H(f)$ be the transfer function of a band-pass filter. For a sinusoidal signal, the bandwidth is defined between the 3-dB points where the output power is decreased to half the maximum value. However, for a stochastic driving signal, the output power is a function of both $H(f)$ and $S(f)$, the latter being the spectrum of the driving signal. This requires a new definition for the bandwidth:

$$B_n = \frac{\displaystyle\int_0^{\infty} |H(f)|^2 S(f)\, df}{|H(f_0)|^2 S(f_0)} \tag{5-91}$$

where f_0 is a reference frequency suitably chosen within the pass-band, e.g., the band-center frequency. According to Fig. 5-16, B_n is the base of a rectangle which has an area equal to the area below the function $|H(f)|^2 S(f)$, which necessarily depends on $|H(f_0)|^2 S(f_0)$ also. This requires an exact definition of f_0 for transfer functions with several extreme values; even international standards allow several definitions.[38,39] The power spectrum of the input noise is also of importance; in the cited standards, white noise is assumed. The author is of the opinion that the definition given in Eq. (5-91) is more general and naturally includes the case of white-noise excitation.

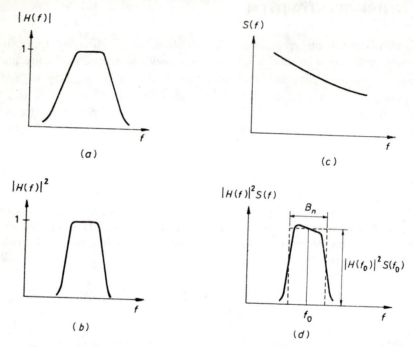

Figure 5-16 *(a)* Transfer characteristic of band-pass filter. *(b)* Squared absolute value of the transfer function. *(c)* Noise power spectrum. *(d)* Determination of the noise bandwidth.

As an example, let us consider the noise bandwidth of a low-pass RC network made up of a series R and a parallel C for a white-noise signal of power density $S(f) = S(f_0) = A$ when $f_0 = 0$, so $H(f_0) = 1$. Thus

$$B_n = \int_0^\infty \frac{1}{1+(f/f_a)^2} \, df = f_a \arctan \frac{f}{f_a} \Big|_0^\infty = \frac{\pi}{2} f_a \qquad (5\text{-}92)$$

where $f_a = 1/2\pi RC$ is the upper cut-off frequency pertaining to the 3-dB attenuation. Table 5-1 presents noise bandwidth values related to the 3-dB bandwidth measurable with a sine-wave signal, as calculated from Eq. (5-91) for amplifiers comprising synchronously tuned resonant circuits and maximally flat band-pass

Table 5-1 Equivalent noise bandwidths

Number of stages m	$B_{noise}/B_{sine\ wave}$	
	Synchronously tuned resonant circuits	Maximally flat band-pass filters
1	1.57	1.57
2	1.22	1.11
3	1.16	1.05
4	1.13	1.03
5	1.11	1.02
6	1.10	1.01
∞	1.06	1.0

filters. These values, as well as the case of $S(f) \cong f^{-2}$ excitation, may be analytically calculated[34] utilizing the known methods of network theory.

A more complicated situation arises for $S(f) \cong f^{-1}$, i.e. in the case of flicker noise.[1] Consider a band-pass filter with rectangular response, having a transfer function of

$$H(f) = \begin{cases} 1 & \text{if} & f_1 < f < f_2 \\ 0 & \text{if} & f < f_1 \quad f > f_2 \end{cases} \tag{5-93}$$

and this has a bandwidth of $B_w = f_2 - f_1$ for both sine-wave and white-noise excitation. However, for flicker-noise excitation we have, according to (5-91),

$$B_f = \frac{\int_{f_1}^{f_2} S(f)\, df}{S(f_0)} = \frac{\int_{f_1}^{f_2} S_c(f_c/f)\, df}{S_c(f_c/f_0)} = f_0 \ln \frac{f_2}{f_1} \tag{5-94}$$

where S_c is the power density extrapolated to the crossover frequency f_c (Fig. 5-17). The ratio referred to the white-noise excitation is

$$\frac{B_f}{B_w} = \frac{f_0}{f_2 - f_1} \ln \frac{f_2}{f_1} \tag{5-95}$$

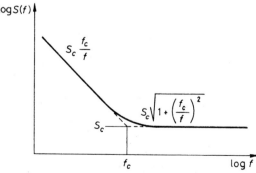

Figure 5-17 Power spectrum of transistor noise.

Table 5-2 Flicker-noise relative bandwidth values for a band-pass filter of rectangular response and bandwidths B_w

$(f_2-f_1)/f_0$	B_f/B_w	
	$f_0 = (f_1+f_2)/2$	$f_0 = \sqrt{f_1 f_2}$
0.2	1.0024	0.9984
0.6	1.0316	0.9843
1.0	1.0986	0.9624

and thus depends on the choice of f_0. Table 5-2 shows values of B_f/B_w for three different relative bandwidths and for two different band-center frequencies. The transition region between the flicker-noise and white-noise ranges for bipolar transistors may be approximated by the expression

$$S(f) = S_c \sqrt{1 + \left(\frac{f_c}{f}\right)^2}$$

(5-96)

For the bandwidth B_c applying this crossover region, the following ratio can be calculated:

$$\frac{B_c}{B_w} = \frac{\sqrt{1+\varphi_2^2} - \sqrt{1+\varphi_1^2} + \ln\left[\varphi_2\left(1+\sqrt{1+\varphi_1^2}\right)/\varphi_1\left(1+\sqrt{1+\varphi_2^2}\right)\right]}{(\varphi_2-\varphi_1)\sqrt{1+(f_c/f_0)^2}}$$

(5-97)

in accordance with Eq. (5-95), where $\varphi_1 = f_1/f_c$ and $\varphi_2 = f_2/f_c$ (see Fig. 5-18). For $\varphi_2 \ll 1$, B_c/B_w will asymptotically approximate B_f/B_w.

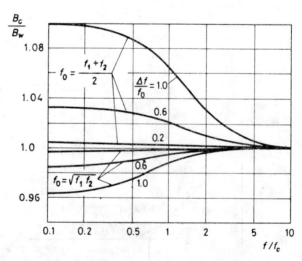

Figure 5-18 Noise bandwidth of filter having a rectangular characteristic in the flicker region and the crossover region related to white noise.

The response of real band-pass filters may be better approximated by the trapezoid form shown in Fig. 5-19 instead of the rectangle. For this response, we have from Eq. (5-91) for white noise

$$B_w = \int_0^{f_1} \left(\frac{f}{f_1}\right)^n df + \int_{f_1}^{f_2} df + \int_{f_2}^{\infty} \left(\frac{f_2}{f}\right)^n df = \frac{f_1}{n+1} + f_2 - f_1 + \frac{f_2}{n-1} \qquad (5\text{-}98)$$

and the expression corresponding to Eq. (5-95), applying to flicker noise, is

$$\frac{B_f}{B_w} = \frac{f_0(2/n + \ln f_2/f_1)(n^2 - 1)}{n[n(f_2 - f_1) + f_2 + f_1]} \qquad (5\text{-}99)$$

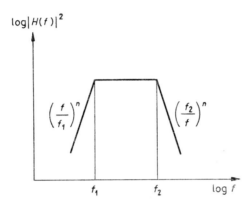

Figure 5-19 Trapezoid approximation of
band-pass filter characteristic.

Even more complicated expressions are derived for the crossover region described by the spectrum [Eq. (5-96)][1] but may be expressed in closed form for integer values of the exponent n. Figure 5-20 shows the functions B_c/B_w for trapezoidal filters with two different slopes, with two different band-center frequencies, and three relative bandwidth values.

B_c/B_w will differ only slightly from unity even at high relative bandwidth values for a correct choice of the band center. This is primarily required in highly accurate noise spectrum analysis. However, the filter characteristic is generally given graphically or in tabulated form. In this case, the band-center value f_0 may be determined from the integral equation

$$\int_0^{f_0} |H(f)|^2 S(f)\, df = \int_{f_0}^{\infty} |H(f)|^2 S(f)\, df \qquad (5\text{-}100)$$

which may be given either graphically or numerically.

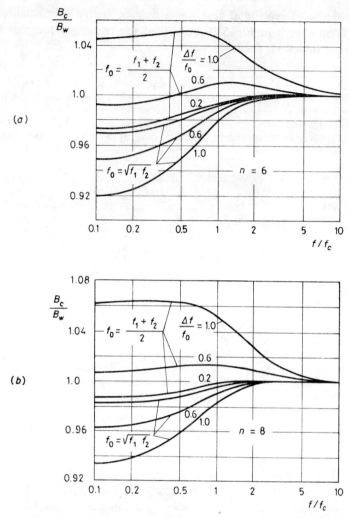

Figure 5-20 Noise bandwidth of filter having trapezoidal characteristic in the flicker region and the crossover region related to white noise. (a) $n=6$. (b) $n=8$.

PROBLEMS

5-1 A load resistance of 100 kΩ is connected in series with a reverse biased p–n junction (Fig. P5-1). The reverse current is 1 μA. What is the noise voltage across this resistance in a bandwidth of 100 kHz?

Answer: The shot-noise current of the diode is

$$i = \sqrt{2qI_0\Delta f} = \sqrt{2\times1.6\times10^{-19}\times10^5} = 1.79\times10^{-10}\,\text{A}$$

As the differential resistance of the diode is infinite in the ideal case, this current fully passes through the load resistance and gives rise to a voltage drop of

$$iR = 1.79\times10^{-10}\times10^5 = 17.9 \ \mu V$$

However, at $T = 300$ K, a thermal-noise voltage of

$$u = \sqrt{4kTR\varDelta f} = \sqrt{4\times1.38\times10^{-23}\times300\times10^5\times10^5} = 13 \ \mu V$$

is simultaneously generated in the resistance. As u and iR are independent of each other, the overall noise voltage, according to Eq. (5-22), is

$$u_r = \sqrt{u^2+(iR)^2} = 22.2 \ \mu V$$

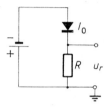

Figure P5-1 Diode
with series resistance.

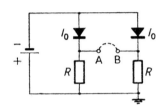

Figure P5-2 Complex
network.

5-2 What is the noise voltage between points A and B of the circuit shown in Fig. P5-2? If A and B are connected, what is the noise voltage across the load resistance?

Answer: Both A and B show a noise voltage of 22.2 μV relative to the earth terminal. The two branches being independent, the overall noise voltage is

$$u_r = \sqrt{u_A^2+u_B^2} = \sqrt{2}\times22.2 = 31.4 \ \mu V$$

In case of parallel connection the overall noise current of the two diodes is

$$i = \sqrt{2q(I_0+I_0) \ \varDelta f} = 2.53\times10^{-10} \ A$$

(again because of independent branches). The voltage drop across the resulting load resistance is

$$iR = 2.53\times10^{-10}\times5\times10^4 = 12.7 \ \mu V$$

The thermal-noise voltage of the resulting resistance is

$$u = 13/\sqrt{2} = 9.2 \ \mu V$$

Finally, the overall noise voltage is

$$u_r = \sqrt{12.7^2+9.2^2} = 15.7 \ \mu V$$

REFERENCES

[1] Ambrózy, A.: "Bandbreitenprobleme bei präziser Rauschmessung von elektronischen Bauelementen", *ATM. Lfg.*, vol. 260, no. 1, pp. 21—24, 1966.

[2] —: "Optimum Noise Matching of Temperature Sensors" (Hőérzékelők optimális zajillesztése), *Híradástechnika*, vol. 28, no. 6, pp. 161—165, 1977 (in Hungarian).

[3] Becking, A. *et al.*: "The Noise Factor of Four Terminal Networks", *Philips Res. Rept.*, vol. 10, no. 5, pp. 345—357, 1955.

[4] Bode, H. W.: *Network Analysis and Feedback Amplifier Design*, D. Van Nostrand, Princeton 1945.

[5] Branin, F. H.: "Network Sensitivity and Noise Analysis Simplified", *IEEE Trans.*, vol. CT-20, no. 3, pp. 285—288, 1973.

[6] Choma, J.: "A Model for the Computer-Aided Noise Analysis of Broad-Banded Bipolar Circuits", *IEEE J.*, vol. SC-9, no. 6, pp. 429—435, 1974.

[7] Dement'ev, E. P.: *Elements of General Theory and Calculation of Noisy Linear Circuits*, Gosenergoizdat, Moscow 1963, (in Russian).

[8] Desoer, C. A. and E. S. Kuh: *Basic Circuit Theory*, McGraw-Hill, New York 1969.

[9] Director, S. W. and R. A. Rohrer: "Interreciprocity and Its Implications", *Proc. Int. Symp. Network Theory*, Belgrade, September, 1968, pp. 11—30.

[10] Faulkner, E. A.: "The Principles of Impedance Optimization and Noise Matching", *J. Phys.*, vol. E8, no. 7, p. 540, 1975.

[11] Feifel, B.: "Das Rauschverhalten gegengekoppelter Verstärker", *AEÜ*, vol. 20, no. 1, pp. 12—18, 1966.

[12] Finnegan, P. J.: "Noise Analysis of Feedback Amplifiers", *Electr. Eng.*, vol. 39, no. 10, pp. 612—616, 1967.

[13] Friis, H. T.: "Noise Figure of Radio Receiver", *Proc. IRE*, vol. 32, no. 7, pp. 419—422, 1944.

[14] Fukui, H.: "Available Power Gain, Noise Figure and Noise Measure of Two-Ports", *IEEE Trans.*, vol. CT-13, no. 2, pp. 137—142, 1966.

[15] Géher, K.: "The Theory of Sensitivity Invariants and Their Application to Optimization of Tolerances and Noises", *Period. Polytechn.*, vol. 19, no. 1, pp. 25—34, 1975.

[16] Hartman, K.: "Noise Characterization of Linear Circuits", *IEEE Trans.*, vol. CAS-23, no. 10, pp. 581—590, 1976.

[17] Haus, H. and R. Adler: *Circuit Theory of Linear Noisy Networks*, John Wiley, New York, 1959.

[18] — *et al.*: "Representation of Noise in Linear Two-Ports", *Proc. IRE*, vol. 48, no. 1, pp. 69—74, 1960.

[19] Heffner, H.: "The Fundamental Noise Limit of Linear Amplifiers", *Proc. IRE*, vol. 50, no. 7, pp. 1604—1608, 1962.

[20] Hillbrand, H. and P. H. Russer: "An Efficient Method for Computer-Aided Noise Analysis of Linear Amplifier Networks", *IEEE Trans.*, vol. CAS-23, no. 4, pp. 235—238, 1976.

[21] Holt, A. G. J. and M. R. Lee: "A Relationship between Sensitivity and Noise", *Int. J. Electr.*, vol. 26, no. 6, pp. 591—594, 1969.

[22] Iversen, S.: "The Effect of Feedback on Noise Figure", *Proc. IEEE*, vol. 63, no. 3, pp. 540—542, 1975.

[23] Kässer, R.: "Optimum Source Admittance Matching for Low-Noise Receiver without Transformer", *Electr. Lett.*, vol. 9, no. 6, pp. 124—125, 1973.

[24] —: "Warum nicht ein Operationsverstärker als rauscharmer Element?" *Technica*, vol. 23, no. 1, pp. 19—23, 1974.

[25] Kleckner, K. R.: "Correlation of Noise Generators", *Proc. IEEE*, vol. 53, no. 2, p. 202, 1965.

[26] Lange, J.: "Noise Characterization of Linear Two-Ports in Terms of Invariant Parameters", *IEEE J.*, vol. SC-2, no. 2, pp. 37—40, 1967.

[27] Mavor, J.: "Noise Parameters for MOS Transistors", *Proc. IEE*, vol. 113, no. 9, pp. 1463—1467, 1966.

[28] Meyer, R. G. *et al.*: "Computer Simulation of $1/f$ Noise Performance of Electronic Circuits", *IEEE J.*, vol. SC-8, no. 3, pp. 237—240, 1973.

[29] Oliver, B. M.: "Thermal and Quantum Noise", *Proc. IEEE*, vol. 53, no. 5, pp. 436—454, 1965.

[30] Roengpithya, V. and A. R. Boothroyd: "Behaviour of Minimum Excess Noise Figure and Maximum Transducer Gain in a Linear Two-Port", *Electr. Lett.*, vol. 2, no. 8, pp. 290—291, 1966.

[31] Rohrer, E. A. *et. al.*: "Computationally Efficient Electronic Circuit Noise Calculations", *IEEE J.*, vol. SC-6, no. 4, pp. 204—213, 1971.

[32] Rothe, H. and W. "Dahlke: Theory of Noise Four-Poles", *Proc. IRE*, vol. 44, no. 6, pp. 811—818, 1956.

[33] Spence, R.: *Linear Active Networks*, Wiley Interscience, New York. 1970.

[34] Trofimenkoff, F. N.: "Noise Margins of Bandpass Filters", *IEEE Trans.*, vol. CT-20, no. 2, pp. 171—172, 1973.

[35] Valkó, I. P.: "Noise Parameters of Acoustical Equipments" (Akusztikai berendezések zajossága), *MTA Műsz. Tud. Oszt. Közl.*, vol. 27, no. 3—4, pp. 313—332, 1960, (in Hungarian).

[36] van der Ziel, A.: *Noise*, Prentice Hall, Englewood Cliffs, 1954.

[37] —: *Noise in Measurements*, John Wiley, New York, 1976.

[38] IEC Standard 225, "Octave, Half-Octave and Third-Octave Band Filters Intended for the Analysis of Sound and Vibration".

[39] IEC Standard 561, "Electro-Acoustical Measuring Equipment for Aircraft Noise Certification".

CHAPTER

SIX

NOISE OF SEMICONDUCTOR DEVICES

We now possess all the methods needed for the thorough investigation and modeling of noise phenomena in actual (active or passive) devices. However, this book covers only the most frequently used semiconductor devices; for other devices, reference is made to the literature.[139, 151]

6-1 THE p–n JUNCTION[38, 112, 137, 155]

In most semiconductor devices, one or more p–n junctions are present. The equilibrium energy band diagram of the junction is shown in Fig. 6-1a, and the current components flowing through the device are depicted in Fig. 6-1b. The majority charge carriers, after passing the potential gap, are taken up into the adjoining layer of opposite conduction type by diffusion, while the minority carriers are accelerated in the reverse direction by the potential gap. In the absence of an external voltage, the sum of the currents due to the majority and minority charge carriers is zero. Without recombination centers in the vicinity of the junction, the equilibrium conditions are independent for both the hole currents and electron currents.

Assuming an ideal p–n junction, the relationship between the overall current and the external voltage is

$$I = I_0(e^{U/U_T} - 1) \tag{6-1}$$

where $U_T = kT/q$ is the thermal voltage, and

$$I_0 = qA \left(p_n \sqrt{\frac{D_p}{\tau_p}} + n_p \sqrt{\frac{D_n}{\tau_n}} \right) \tag{6-2}$$

Here A is the cross section, p_n and n_p are the minority carrier densities, D_p and

D_n are diffusion constants, and τ_p and τ_n are the charge carrier life times due to direct recombination.

An important result is seen by a small formal rearrangement of Eq. (6-1).

$$I = I_0 e^{U/U_T} - I_0 = (I+I_0) - I_0$$
$$= \text{majority current} - \text{minority current} \tag{6-3}$$

Both the majority and the minority current components have shot noise, as there is a random fluctuation in charge carriers passing through the junction. However, the two components are independent of each other, so at low frequencies, as a consequence of Eqs (1-1) and (5-22), we have

$$\overline{i^2} = 2q[(I+I_0)+I_0]\Delta f = 2q(I+2I_0)\Delta f \tag{6-4}$$

An interesting result

$$\overline{i^2} = 4qI_0\Delta f \tag{6-5}$$

is obtained if $I=0$: and the differential conductance of the junction is

$$G = \frac{dI}{dU} = \frac{q}{kT}(I+I_0)$$

$$G_{I=0} = \frac{qI_0}{kT} \tag{6-6}$$

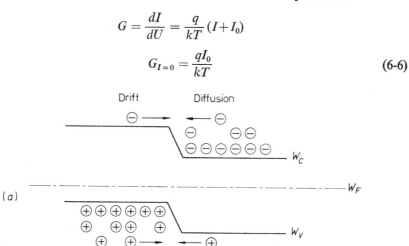

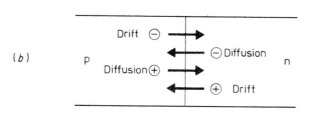

Figure 6-1 The p–n junction in the equilibrium state. *(a)* Energy band diagram. *(b)* Current components.

Comparing Eqs. (6-5) and (6-6), we have

$$i^2 = 4kTG_{I=0}\Delta f \qquad (6\text{-}7)$$

i.e. the p–n junction at equilibrium acts similarly to a homogeneous resistance not only from the small-signal aspect but also from the noise aspect.

The value of i^2 is affected by the complex admittance of the junction at much lower frequencies than the transit-time effect becomes noticeable according to Eq. (4-20); Eq. (6-4) must be modified accordingly.

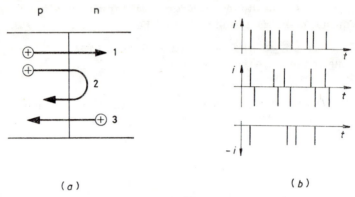

(a)   (b)

Figure 6-2 *(a)* Hole currents in the p–n junction: 1. diffusion component, 2. reversing diffusion component, and 3. drift component. *(b)* Pulses pertaining to the passage of elementary charges. (*Courtesy of Butterworth, London.*[141])

Figure 6-2a is an interpretation of the corpuscular model. Only the motion of holes will be investigated. Three cases can be distinguished:

1. Majority holes are carried by diffusion from the p side to the n side.
2. The hole coming from the p side penetrates into the n side and then returns to the p side.
3. The minority holes in the n side drift into the p side by the field within the junction.

Components 1 and 3 have already been taken into account in Eqs. (6-3) and (6-4). Component 2 does not introduce stationary current in the external circuit. Transition of a charge carrier through the p–n junction generates an approximate Dirac pulse in the external circuit [see Sec. 4-1 and Eq. (4-20) for the power spectrum]. According to Fig. 6-2b, component 2 is given by a series of alternate positive and negative pulses.

Let us assume that all holes remain for the same time τ in the n side. The time function of component 2 would then be, because of the phase delay $\omega\tau$, of the form

$$i_2 = I_2 - I_2 e^{-j\omega\tau} \qquad (6\text{-}8)$$

From this relation, the junction admittance due to i_2 may be calculated. Thus

$$Y_2 = G_2 + jB_2 = \frac{di_2}{du} = \frac{qI_2}{kT}(1 - e^{-j\omega\tau}) \tag{6-9}$$

Y_2 disappears when $\omega = 0$, but in the high-frequency range, $G_2 \neq 0$ and $B_2 \neq 0$. Taking into account that the holes have different storage times τ, following the density function $g(\tau)$ not yet defined, G_2 may be written as

$$G_2 = \text{Re}(Y_2) = \int_0^\infty \frac{qI_2}{kT}(1 - \cos \omega\tau) g(\tau) \, d\tau \tag{6-10}$$

B_2 may be derived in a similar way.

The subsequent positive and negative Dirac pulses are perfectly correlated during the storage time τ, but the pulse pairs are independent of each other. Taking this into account, the mean square value of the I_2 current fluctuation for $\tau = \text{const.}$ is

$$i_2^2 = 2qI_2\Delta f |1 - e^{-j\omega\tau}|^2 = 4qI_2\Delta f(1 - \cos \omega\tau) \tag{6-11}$$

Again taking into account the density function $g(\tau)$, we have

$$i_2^2 = \int_0^\infty 4qI_2\Delta f(1 - \cos \omega\tau) g(\tau) \, d\tau = 4kTG_2\Delta f \tag{6-12}$$

where Eq. (6-10) has also been considered. This means that noise current i_2 may be considered to be the thermal-noise current of $\text{Re}(Y_2)$. This, of course, is valid for all systems that are in thermodynamic equilibrium with their environments. The holes generating current component 2 return to their starting point and thus do not consume energy from the circuit.

The overall admittance of the p–n junction is determined by all three components shown in Fig. 6-2.

$$Y = G + jB = G_1 + G_2 + G_3 + j(B_1 + B_2 + B_3) \tag{6-13}$$

$$G_1 = qI_1/kT \quad G_3 = 0 \tag{6-14}$$

$$B = B_2 + j\omega C_T \tag{6-15}$$

where C_T is the space charge capacity of the junction. Thus it follows that

$$G_2 = G - G_1 \tag{6-16}$$

and the complete squared noise current, utilizing Eqs (6-4) and (6-12), is

$$i^2 = i_1^2 + i_2^2 + i_3^2 = 2q(I + I_0)\Delta f + 4kT(G - G_1)\Delta f + 2qI_0\,\Delta f \tag{6-17}$$

This may be simplified by substituting Eq. (6-6). Thus

$$\boxed{i^2 = 4kTG\Delta f - 2qI\Delta f} \tag{6-18}$$

This formula gives the true internal noise current of the p–n junction for all frequencies and working points. Naturally, the noise current flowing through the external circuit may again be calculated from Eq. (4-21). It is noted that other methods are also known for the derivation of Eq. (6-18).[16,141]

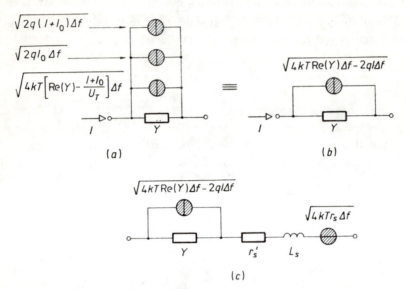

Figure 6-3 Noise equivalent circuits of the p–n junction. *(a)* According to Eq. (6-17). *(b)* According to Eq. (6-18). *(c)* Noise equivalent circuit of semiconductor diode.

Utilizing Eq. (6-17), the noise equivalent circuit of the p–n junction may be drawn as shown in Fig. 6-3a. The differential admittance of the junction at an arbitrary frequency due to the charge storage effect[89] is

$$Y = G_1(1+j\omega\tau_p)^{1/2}+j\omega C_T \qquad (6\text{-}19)$$

Contracting the first and last terms of Eq. (6-17), we obtain the shot-noise components [see Eq. (6-4)]. Thus

$$i_{shot}^2 = 2q(I+2I_0)\varDelta f \qquad (6\text{-}20)$$

The center term may be considered to have a thermal origin. Thus

$$i_{therm}^2 = 4kT(G-G_1)\varDelta f = 4kT(\text{Re } Y - G_1)\varDelta f \qquad (6\text{-}21)$$

This means that the charge storage effect generates not only additional admittance but also noise. At low frequencies where $\omega\tau_p \ll 1$, expression (6-19) in an expanded form may be substituted into Eq. (6-21), resulting in

$$i_{therm}^2 \cong 4kTG_1\frac{\omega^2\tau_p^2}{8}\varDelta f \qquad (6\text{-}22)$$

6-2 DIODES

A perfect realization of a p–n junction is not possible. A simple diode comprises, in addition to the p–n junction at least a series resistance r_s. The equivalent circuit shown in Fig. 6-3a may be expanded to include this resistance, and the thermal-noise voltage $\sqrt{4kTr_s\Delta f}$ of the resistance may be taken into account.

However, the series resistance is not constant but depends on the current flowing through the resistance because of the ambipolar diffusion.[150] Generally

$$r_s = r_{s0} + I\frac{\partial r_s}{\partial I} \tag{6-23}$$

where $\partial r_s/\partial I < 0$. This is caused by an increase in the charge carrier density and thus a decrease in the specific resistance in the layer of lower doping due to the increase in current. At higher frequencies, $\partial r_s/\partial I$ becomes complex,[127] modifying Eq. (6-23). Thus

$$Z_s = r_{s0} + I\frac{\partial r_s}{\partial I}(\omega) = r_s' + j\omega L_s \tag{6-24}$$

The noise equivalent circuit will be as shown in Fig. 6-3c; here i^2 may be calculated from Eq. (6-18), and

$$u^2 = 4kT\,\mathrm{Re}\,(Z_s)\Delta f \tag{6-25}$$

During the derivation of our previous relations direct recombination has been assumed. Imperfect p–n junctions comprise recombination centers, so it may happen that the hole recombines before reaching the n side but subsequently escapes from the trap. The recombination–generation current in the conducting direction of the diode is given by these holes and electrons, respectively.[118]

The current–voltage characteristic of an ideal p–n junction is

$$I = I_0(e^{U/U_T} - 1) \tag{6-26}$$

while that of a pure recombination current is

$$I_R = I_{0R}(e^{U/2U_T} - 1) \tag{6-27}$$

The sum of the two currents flows through the actual diode. The current–voltage characteristic of the actual diode is preferably approximated by

$$I \cong I_0(e^{U/mU_T} - 1) \tag{6-28}$$

Exact calculation has shown[103] that the factor two in the exponent denominator of Eq. (6-27) is only valid for generation–recombination centers located precisely in the middle of the depletion layer. For other cases, any value in the range $1 < x < 2$ is possible, and in Eq. (6-28), m will be correspondingly modified too.

Considering the noise due to the generation–recombination current component, the most simple theory assumes that the passage of the charge carriers through the depletion layer is a two-phase process, due to a resting time interval in the generation–recombination center. An elementary part of this two-phase process may be considered to be the passage of charge $q/2$. Assuming a generation–recombination current I_R, the squared fluctuation of the overall hole plus electron current is

$$i_{RG}^2 = i_p^2 + i_n^2 = 2\frac{q}{2}I_R\Delta f = qI_R\Delta f = 2qI_R F_{RG}^2 \Delta f \qquad (6\text{-}29)$$

where the noise-attenuating factor F_{RG}^2 has been applied according to the terminology introduced in Sec. 4-1.

The real situation is considerably more complicated. Taking into account the distribution of the recombination time constants, we have $0.75 < F_{RG}^2 < 1,$[82] which shows instantly that the determination of the noise-attenuating factor is not a simple task. Some attempts[166, 168] have produced values between 0.8 and 0.85. Recent experimental results have paved the way for theoretical investigations.[159, 161] The noise-attenuating factor is probably 0.75 at low frequency and varies between 0.5 and 1.0 in the high-frequency range; the smaller value applies for symmetrical junctions ($N_A = N_D$), the larger number for highly asymmetrical junctions.

In the case of diodes operating at high injection levels, the ambipolar diffusion effect has also to be taken into account, in addition to the generation–recombination noise. This may be accomplished using a transmission line model.[157]

The recombination may be neglected in reverse biased diodes as the minority charge carriers pass through the junction at high speed.

In the case of silicon diodes having a wide depletion layer, a significant part of the reverse current is due to the charge carrier pairs generated in the depletion layer. However, these pairs are separated immediately after their generation, so a single hole or electron does not pass the complete depletion layer, in contrast to the minority carriers generated in the neutral zones. It may be shown[121] that from the noise aspect a factor of 2/3 has to be taken into account for the generation component, i.e.

$$i^2 = i_0^2 + i_{gen}^2 = 2qI_0\Delta f + \frac{2}{3}2qI_{gen}\Delta f \qquad (6\text{-}30)$$

where i_{gen}^2 is the squared fluctuation in the generation current I_{gen}. Investigation of the fine structure of the depletion layer has resulted in a multiplying factor slightly different from 2/3.[147]

It is well known for diodes that are small or have poor environmental thermal coupling that thermal instability may take place for high reverse voltages. This instability is the consequence of a positive feedback generated thermally. The same kind of feedback also increases the diode noise, at least in the range stated.[149]

The noise phenomena due to the avalanche breakdown effect have already been covered in Sec. 4-4.

Flicker noise may be generated in both forward and reverse directions. In the forward direction, flicker noise may be generated by the slow surface states or by the recombination centers near the surface modulated by the oxide charges (see Sec. 4-6).[37,51,61,85] According to some theories,[96,97] the primary cause is white noise which is amplified by positive thermal feedback, and the $1/f$ spectrum thus produced is a consequence of the diode structure which may be described by distributed thermal parameters.

Flicker noise in reverse biased diodes generally indicates a poor junction.[50] It is shown in Fig. 6-4 that, in the layer of a lower doping, an inversion channel may be generated due to surface conditions, leading to an increase in the reverse current and flicker noise.

The burst noise treated in Sec. 4-7 has also been observed in p–n junctions. We refer the reader to the appropriate items in the references of Chapter 4.

The technological development of Schottky diodes and their widening applications have resulted in their noise investigations too. The energy diagram of the Schottky diode is shown in Fig. 6-5. Here too, it may be expected that the current–voltage characteristic is an exponential function. However, the external voltage applied to the junction can influence not only the height but, to a lesser extent, the shape of the barrier. The characteristic should therefore be always expressed in the form

$$I = I_0(e^{U/mU_T} - 1) \qquad (6\text{-}31)$$

where $m > 1$ is called the "ideality factor".

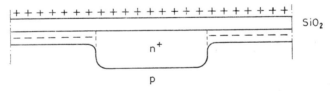

Figure 6-4 Inversion channel due to surface impurities.

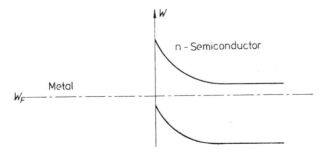

Figure 6-5 Energy diagram of Schottky barrier.

According to experimental results,[22, 100, 101] Eq. (6-18) may be applied here too. Thus

$$i^2 = \left(4kTG - \frac{2qI}{m}\right)\Delta f \qquad (6\text{-}32)$$

Now there is no danger of an abrupt change in G with an increase in frequency; and as the minority charge carrier plays no role in the Schottky diodes, there is no diffusion admittance either. However, Eq. (6-32) is generally only valid at room temperature. If the diode is cooled, the original thermal emission responsible for the diode function changes to mixed thermal/space emission, and finally to pure space emission. For this case, Eq. (6-32) should be modified.[164]

Unfortunately, the Schottky diode has additional noise sources. For example, if there are generation–recombination centers in the depletion layer, these may communicate with the metal and the n-type semiconductor, respectively, by a two-phase tunnel effect. The occupied state of the generation–recombination centers modulates the barrier height, and the current is modulated (exponentially) by the latter. Taking into account the extremely wide time constant range due to the tunnel effect [see Eqs. (4-97) to (4-100)], the generation of $1/f$ noise may be expected.[52, 55]

The current injected by the metallic edge of a diode may also be a source of substantial flicker noise.[169] There is a tendency to decrease this current component utilizing refined technology because this current has an adverse effect on the reverse stationary characteristic too. A recent analysis showed that mobility fluctuations and series resistance should also be taken into consideration.[76]

6-3 BIPOLAR TRANSISTORS[9, 138, 140]

A simple, four-element equivalent circuit of the bipolar transistor valid for small-signal operation is shown in Fig. 6-6.[89] The differential resistance of the forward biased emitter junction is $r_e = U_T/I_E$; that of the collector junction is $r_c < \infty$ due to the Early effect, the charge carrier generation in the depletion layer, and the imperfections on the surface. The current-controlled current generator represents the connection between collector and emitter currents, but the effect of the collector side on the emitter in the reverse direction is neglected. The resistance of the base lead is $r_{bb'}$.

In Fig. 6-6, the noise of the p–n junctions is modeled by the parallel-connected current generators i_1 and i_2, and the thermal noise of the base resistance is modeled by the voltage generator u_b. The emitter junction is biased in the forward direction, so as in Eq. (6-4), we have

$$i_1^2 = 2q(I_E + I_{E0})\Delta f + 2qI_{E0}\Delta f \cong 2qI_E\Delta f \qquad (6\text{-}33)$$

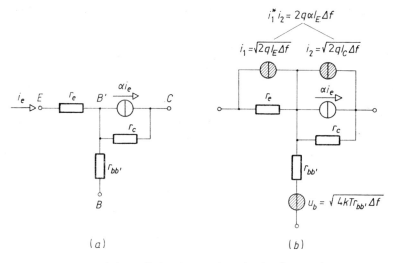

Figure 6-6 *(a)* Small-signal equivalent circuit of a transistor.
(b) Extension by noise sources.

where the first and second terms originate from the current fluctuation of the majority and minority charge carriers, respectively. (The latter is normally not to be considered as it is negligible in modern semiconductor devices.) Only minority current flows in the reverse biased collector junction, so

$$i_2^2 = 2qI_C\Delta f \tag{6-34}$$

As the collector current is not independent of the emitter current, there is a strict correlation between i_1 and i_2. Thus

$$i_1^* i_2 = i_1^* \alpha i_1 = \alpha |i_1|^2 \cong 2q\alpha I_E\Delta f = 2kTY_{21}\Delta f \tag{6-35}$$

However, Eq. (6-35) is true only at low frequencies. At higher frequencies, the diffusion process treated in Eqs. (6-8) to (6-17) must also to be taken into account as this has an effect both on the signal amplification of the transistor and on the relationship between i_1 and i_2. If $i_2 = g_m u_{eb}$ applies for low frequencies, at high frequencies this has to be modified for $i_2 = Y_{21} u_{eb}$. Now Eq. (6-35) can also be modified in order to apply for all frequencies. Thus

$$i_1^* i_2 = \frac{Y_{21}}{g_m} 2q\alpha I_E\Delta f = 2kT\alpha Y_e\Delta f \tag{6-36}$$

where the equations $\alpha q I_E/kT = g_m$ and $Y_{21} = \alpha Y_e$ have been utilized.[155]
Finally, the thermal noise of the base resistance is

$$u_b^2 = 4kTr_{bb'}\Delta f \tag{6-37}$$

The equivalent circuit of Fig. 6-6b is not convenient because of the correlated generators, and anyway, it is only valid for low frequencies. Similarly to the reasoning applied for the p–n junction, the emitter junction at high frequencies will be characterized by the relationship

$$\boxed{i_1^2 = [2qI_E + 4kT(G_e - G_{e0})]\,\varDelta f}$$ (6-38)

In this relationship, which is similar to (6-17), G_e is the real part of the frequency-dependent emitter admittance and G_{e0} is the low-frequency value thereof. Equation (6-34) applies to the collector junction.

In order to simplify the equivalent circuit, let us separate the collector noise current i_2 into two components, one in perfect correlation with the emitter current and the other component i_2' independent of the emitter current. Thus

$$i_2 = \alpha i_1 + i_2'$$ (6-39)

i_2' may be calculated with the aid of Eq. (5-20):

$$|i_2'|^2 = i_2' i_2'^* = (i_2 - \alpha i_1)(i_2^* - \alpha^* i_1^*)$$
$$= |i_2|^2 + |\alpha|^2 |i_1|^2 - 2\,\mathrm{Re}\,(\alpha^* i_1^* i_2)$$ (6-40)

Substituting Eqs (6-34), (6-36), and (6-38), we have

$$i_2'^2 = 2qI_C \varDelta f + |\alpha|^2 2qI_E \varDelta f + |\alpha|^2 4kT(G_e - G_{e0})\varDelta f$$

$$-4kT\varDelta f\,\mathrm{Re}\,(\alpha\alpha^* Y_e) = 2q\varDelta f\left[I_C + |\alpha|^2 I_E - |\alpha|^2 \frac{2kTG_{e0}}{q}\right]$$ (6-41)

Introducing the current gain factor $A = I_C/I_E$ which is valid for large-signal (DC) operation of a common emitter circuit, Eq. (6-41) may be further transformed

$$\boxed{i_2'^2 = 2qI_E\varDelta f\left[A + |\alpha|^2 - |\alpha|^2 \frac{2kTG_{e0}}{qI_E} + \left(\frac{I_{CB0}}{I_E}\right)\right]}$$ (6-42)

An additional term representing the shot noise due to the collector–base cut-off current I_{CB0} is also introduced; this may be neglected generally, except in the case of devices operating at high temperatures, and when considering Ge transistors. Figure 6-7 shows the equivalent circuit drawn with the aid of Eqs. (6-34), (6-37), (6-38), and (6-42).

The above relationships and equivalent circuits are generally valid and are applicable for several specific cases.

Low Frequency, Average Working Point

At low frequencies, for average working point current, $\alpha \cong A$ and $kTG_{e0}/qI_E = 1$. For this case, Eq. (6-42) is modified as follows:

$$i_2'^2 = 2qI_E A(1 - A)\varDelta f = 2qAI_B\varDelta f$$ (6-43)

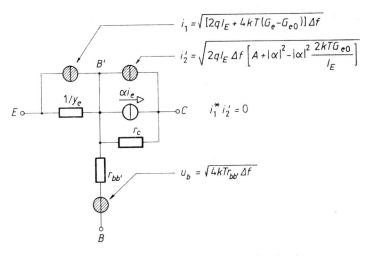

$$i_1 = \sqrt{[2qI_E + 4kT(G_e - G_{e0})]\,\Delta f}$$

$$i_2' = \sqrt{2qI_E\,\Delta f \left[A + |\alpha|^2 - |\alpha|^2\,\frac{2kTG_{e0}}{I_E}\right]}$$

$$i_1^* \, i_2' = 0$$

$$u_b = \sqrt{4kT r_{bb'}\,\Delta f}$$

Figure 6-7 Equivalent circuit comprising uncorrelated noise sources.

Multiplying the above product by I_E/I_E, we have

$$2qAI_B\,\Delta f = 2q\,\frac{I_B I_C}{I_E}\,\Delta f = 2q\,\frac{I_B I_C}{I_B + I_C}\,\Delta f \tag{6-44}$$

which, according to Eq. (4-32), is the squared fluctuation due to the random distribution of currents I_B and I_C.

Low Frequency, Low Current Working Point

At low frequencies, for low working point current[17] or at low temperatures,[14] the recombination in the emitter–base junction is generally significant, especially for Si transistors. The emitter current then is made up of two parts: the diffusion component and the recombination component [see Eqs. (6-26) and (6-27)].

Thus
$$I_E = I_{D0}e^{U/U_T} + I_{R0}e^{U/2U_T} = I_D + I_R \tag{6-45}$$

From this equation, the low-frequency emitter conductance is

$$G_{e0} = \frac{\delta I_E}{\delta U} = \frac{I_D}{U_T} + \frac{I_R}{2U_T} = \frac{1}{U_T}\left(I_D + I_R - \frac{I_R}{2}\right) = \frac{I_E - I_R/2}{U_T} \tag{6-46}$$

Let us eliminate I_R utilizing identities $g_m = \alpha_0 G_{e0} = I_C/U_T$ and the definition $A = I_C/I_E$

$$I_R = \frac{(\alpha_0 - A)2I_E}{\alpha_0} \tag{6-47}$$

Substituting Eqs. (6-46) and (6-47) into (6-42) and rearranging, we have

$$i_2'^2 = 2qI_E \Delta f\left(A + |\alpha|^2 - 2|\alpha|^2 \frac{A}{\alpha_0}\right) \qquad (6\text{-}48)$$

which is frequently used in the form

$$i_2'^2 = 2qI_E \Delta f[A(1-A) + (\alpha_0 - A)^2] \qquad (6\text{-}49)$$

if $\alpha = \alpha_0$. It can be seen that Eq. (6-49) has been extended compared with Eq. (6-43); in this case, the noise is called "apparent" partition noise. The second term in the bracket is only significant if the difference between α_0 and A is high. Let us take into account the following definition of α_0:

$$\frac{\delta I_C}{\delta I_E} = \alpha_0 = A + I_E \frac{\delta A}{\delta I_E} \qquad (6\text{-}50)$$

It can be immediately seen that only a current-dependent A will introduce a difference between A and α_0. This may be the consequence of the recombination or, as will be shown, the high injection level.

At low temperatures, another noise due to generation–recombination may appear: the fluctuation caused by the random occupancy of the generation–recombination centers in the base.[167,168]

High Frequency

At high frequencies, if recombination effects do not have to be taken into account, Eq. (6-42) can be simplified to

$$i_2'^2 = 2qI_E \Delta f(A - |\alpha|^2) \qquad (6\text{-}51)$$

However, in this case

$$|\alpha|^2 = \frac{\alpha_0^2}{1 + (f/f_\alpha)^2} \qquad (6\text{-}52)$$

where α_0 is the low-frequency value of the small-signal common base circuit and f_α is the frequency at which $|\alpha| = \alpha_0/\sqrt{2}$. Substituting $I_C = AI_E$, we have

$$i_2'^2 = 2qI_C\left[1 - \frac{\alpha_0^2}{A}\frac{1}{1 + (f/f_\alpha)^2}\right]\Delta f \qquad (6\text{-}53)$$

For $f \to \infty$, $i_2'^2 \to 2qI_C \Delta f$ [see Eq. (6-34)]. This has the following physical meaning: the correlation between the collector circuit and emitter circuit noise generators, according to Eq. (6-36) becomes weaker at higher frequencies, thus the gain decreases, so the emitter circuit noise becomes less perceptible in the collector circuit.

UHF and Microwave Region

The noise equivalent circuit remains basically unchanged in the UHF and micro-wave region.[104] However, the lead inductances, their mutual inductance, and capacitance should be taken into account.[33,91] Also, a substantial frequency dependence is caused by the unavoidable capacitance C_{bc} of the mesa and planar structures. Figure 6-8 shows the equivalent circuit applicable to this region.[135] Its computation is extremely complicated, so either an analog model is feasible for evaluation[6] or an efficient network analysis program should be used. Alternatively, substantial simplifications may be applied.[7]

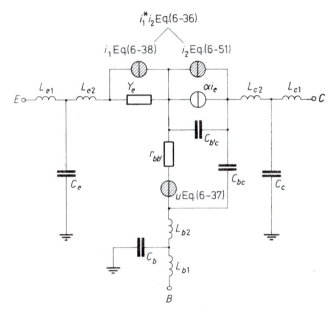

Figure 6-8 Noise equivalent circuit of a transistor in the UHF range.[135] (*Courtesy of the Institute of Electrical and Electtonics Engineers*)

High Injection Levels

The noise relationships at high injection levels are still not completely clarified. This mode of operation is of interest not only for power transistors. High-frequency transistors, due to their small dimensions, may operate under these circumstances even at average working point currents.

The effects due to base modulation have been taken into account when drawing the first equivalent circuit by including R_2 and L_2 (see Fig. 6-9 and the reason-

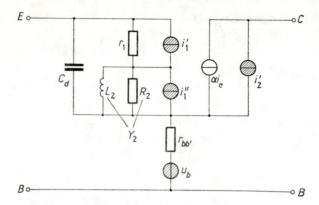

Figure 6-9 Large signal high-frequency equivalent circuit of
a transistor.[120] (*Courtesy of the Institute of Electrical and
Electronics Engineers*)

ing presented at the beginning of Sec. 6-2). The noise generator parameters in
this equivalent circuit are

$$i_1'^2 = 4kT\frac{1}{r_1}\varDelta f - 2qI_E\varDelta f$$

$$i_1''^2 = 4kT\,\text{Re}\,(Y_2)\,\varDelta f$$

$$i_1 = \frac{\alpha}{\alpha_0}Y_{11}(i_1'r_1 + i_1''Z_2) \tag{6-54}$$

$$i_2^2 = 2qI_C\varDelta f$$

$$i_1^*i_2 = 2kT(Y_{21} - Y_{12})\varDelta f$$

Several attempts have been made to verify the above theoretical results[63,136] and
to generalize these results.[95] None of these attempts have produced conclusive,
or even easily applicable, results, so recent efforts[153,165] have been directed at
presenting the approximations valid for extremely high injection levels, omitting the
extremely complicated region of medium injection levels.

It is well known that due to the ambipolar diffusion effect, the equation for
a transistor operating at high levels comprises $U_{EB}/2U_T$ instead of U_{EB}/U_T, in
a similar way to the expression for the recombination current. Under these circumstances, it is feasible to apply a computation sequence similar to Eqs. (6-45) to
(6-49). However, it should also be borne in mind that, in this region, the charge
carriers pass from the base to the emitter by diffusion; thus reducing the emitter
efficiency has an extremely large effect. For p–n–p transistors, the following end

result is obtained with the omission of the details:[153]

$$i_1^2 = i_{1p}^2 + i_{1n}^2 = 2q\Delta f\left[b'I_{Ep}\left(\frac{2G_{ep}}{G_{e0p}}-1\right)+I_{En}\left(\frac{2G_{en}}{G_{e0n}}-1\right)\right] \qquad (6\text{-}55)$$

$$i_2'^2 = 2qI_C\Delta f\frac{b'(1-\alpha_0+f^2/f_\alpha^2)+\alpha_0^2(1-A)/A}{1+f^2/f_\alpha^2} \qquad (6\text{-}56)$$

where the notations have been used as in Eq. (6-42). Subscripts p and n means hole and electron current, respectively, and $b'=(\mu_n+\mu_p)/2\mu_n$.

Common Emitter Equivalent Circuit

Before investigating the additional (1/f and burst) noises of the transistor, it is feasible to change over to the common emitter equivalent circuit. At low frequencies, this circuit is applied almost exclusively. To understand this circuit, Fig. 6-10 should be compared with Fig. 6-6b.[145]

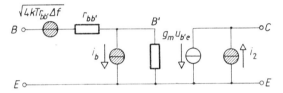

Figure 6-10 Common emitter noise equivalent circuit.

It is evident that the DC current $I_B=I_E-I_C$ and the noise current i_1-i_2 should flow through the base lead leading to node B' in Fig. 6-6b. The base-noise current i_b may be computed in the known way:

$$i_b^2 = (i_1-i_2)(i_1^*-i_2^*) = i_1^2+i_2^2-2\,\mathrm{Re}\,(i_1^*i_2) \qquad (6\text{-}57)$$

Substituting Eqs. (6-34), (6-36), and (6-38), we have

$$i_b^2 = 2\Delta f[q(I_E+I_C)+2kT(G_e-G_{e0})-2kT\,\mathrm{Re}\,Y_{21}] \qquad (6\text{-}58)$$

Taking into account that $I_B=I_E+I_C-2I_C$ and $I_C=g_mkT/q$, the following symmetrical expression is derived:

$$i_b^2 = 2\Delta f[qI_B+2kT(G_e-G_{e0})+2kT(g_m-\mathrm{Re}\,Y_{21})] \qquad (6\text{-}59)$$

Now the only problem to be cleared up is whether a correlation exists between i_b and i_2.

$$i_b^*i_2 = (i_1^*-i_2^*)i_2 = i_1^*i_2-|i_2|^2$$
$$= 2kTY_{21}\Delta f-2qI_C\Delta f = 2kT(Y_{21}-g_m) \qquad (6\text{-}60)$$

At low frequencies, $G_e=G_{e0}$ and $g_m\cong\mathrm{Re}\,Y_{21}\cong Y_{21}$, so Eqs. (6-59) and (6-60) simplify to

$$i_b^2 = 2qI_B\Delta f$$
$$i_b^*i_2 = 0 \qquad (6\text{-}61)$$

Flicker Noise

The exact physical background of the flicker noise is not yet known (see Sec. 4-6); however, the originating site of this noise within the transistor may be fairly well localized. Utilizing the equivalent circuits shown in Figs. 6-6b and 6-10, i_1, i_2, $i_1^* i_2$, and i_b have been measured in the flicker range.[36, 106] Subtracting the white-noise components showed that only i_1 (or i_b) is significant, i.e. the flicker-noise sources are located at the emitter–base junction. Several experiments,[19, 60, 78] illustrated in Fig. 6-11, have been performed in order to make clear whether surface or volume components are dominant.

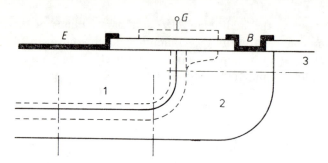

Figure 6-11 Zone division of emitter–base junction in a planar transistor (—·——·—); applied gate electrode for investigation of surface effects (————).

In Fig. 6-11, three regions are shown. In region 1, only the effect of imperfections within the base–emitter depletion layer (e.g. recombination centers) is perceptible. Region 2 is the inactive part of the base resistance (the base lead), and across this resistance a voltage drop may be produced by eventual noise currents. The behavior of region 3 is primarily governed by surface effects. For instance, in the case of a substantial positive charge on the surface, inversion may take place in the base surface region (as shown in Fig. 6-11). In order to investigate the fractional distribution of these surface effects, a control electrode has been positioned on top of the insulating oxide layer above the emitter–base junction. Both surface effects and volume effects have been observed, which may be taken into account by the model shown in Fig. 6-12. This corresponds to the model shown in Fig. 6-10 with the omission of the white-noise sources. Earlier, surface effects have been dominant,[69] but both effects may be equally significant in modern transistors. On the other hand, it has become quite clear that the more perfect the substrate material and the more pure the technology for the transistor production the less will be the flicker noise of the device.[34, 54, 130] However, the perfect structure should be safeguarded during operation of the transistor and care should be taken to avoid improper operating parameters. This is important because the originally "frozen" imperfections may be rearranged by an avalanche breakdown within the emitter–base junction.[94]

A unique explanation of the transistor flicker-noise $1/f$ spectrum under rigorous and universal conditions has not yet been achieved. However, there are some explanations of the working point dependence. It has been shown[25] that the surface and volume effects may be summarized in the expression

$$i_{bf}^2 = 4kT \frac{\varrho_0}{f} \Delta f g_{b'e}^2 \tag{6-62}$$

where ϱ_0 is a constant characteristic of the transistor sample in question and $g_{b'e}$ is the input conductance of the common emitter equivalent circuit. The latter's value may not be calculated in a simple manner because of generation–recombination effects. Let us start from equations similar to (6-26) and (6-28):

$$I_C = I_{C0} e^{U_{BE}/U_T},$$

$$I_B = I_{B0} e^{U_{BE}/mU_T},$$

where $1 \leq m < 2$ because of recombination. From this,

$$B = \frac{I_C}{I_B} = \text{const. } I_C^{1-1/m} \tag{6-63}$$

and

$$\beta = \frac{\partial I_C}{\partial I_B} = \frac{\partial I_C}{\partial U_{BE}} \frac{\partial U_{BE}}{\partial I_B} = \frac{I_C}{U_T} \frac{mU_T}{I_B} = mB \tag{6-64}$$

Utilizing the definition $g_{b'e} = g_m/\beta$, we have

$$g_{b'e} = \frac{I_C}{U_T} \frac{c_1}{I_C^{1-1/m}} = c_2 I_C^{1/m} = c_3 I_B \tag{6-65}$$

Consequently, the noise current in the base circuit i_{bf} is directly proportional to the base DC current. To be more precise, i_{bf} will increase somewhat more slowly than linearly as a function of the collector current I_C. It is interesting to note that a similar conclusion has been reached during a relatively early investigation.[93] However, there is evidence in the literature that i_{bf} is a power function of I_B with an exponent which is greater than one. (See, for instance, the work of Knott;[78] in this case, however, the statement applies only to the surface component.)

Burst Noise

Concerning the pulse-type noise in transistors, reference should be made to Sec. 4-7 as the source of this noise has also been localized in the forward biased emitter–base junction,[60] and the possible noise sources have been analyzed in this Section. The distribution of structural imperfections causing pulse-type noise is probably similar to that of the imperfections giving rise to flicker noise, with the exception of the order of magnitude. It is probably true that flicker noise is caused by submicroscopic imperfections, while burst noise is effected by microscopic imperfections.

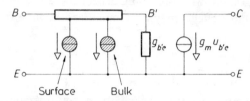

Figure 6-12 Equivalent circuit for showing flicker-noise sources of planar transistor. (*Courtesy of the Institute of Electrical and Electronics Engineers.*)

Because of their similar placement, Figs. 6-11 and 6-12, originally devised for flicker noise, may be used here too. For example, Fig. 6-12 may be extended by a third noise-current generator connected to a chosen point on the base resistance.[21] According to measurements,[60] the current pulse amplitude due to the equivalent generator is

$$i = \text{const.} \; T^{3/2} e^{-(W_g - qU_{BE})/2kT} e^{-T/T_0} \tag{6-66}$$

The frequency of switch-over [v_1 and v_2 in Eqs. (4-108)] depends on the emitter current density,[79, 80] and is a steep function of the temperature.[21]

6-4 NOISE FIGURE OF BIPOLAR TRANSISTORS[102]

The noise equivalent circuits above are suitable for calculating any noise characteristic by applying the methods shown in Chapter 5. The determination of the two-generator equivalent circuit and of the noise figure has practical significance and thus deserves separate attention.

The two-generator equivalent circuit shown in Fig. 5-6 may be determined by starting from Fig. 6-7. However, it is advisable to draw the Thévenin equivalent circuit of the emitter circuit one-port (see Fig. 6-13a). Here

$$u_1^2 = [2qI_E + 4kT(G_e - G_{e0})]\Delta f r_e^2 = 2kT r_e \Delta f \tag{6-67}$$

if $G_e = 1/r_e$ (at not too high frequencies). The source current of the current generator shown in Fig. 5-6 may be most easily calculated with the terminations shown in Fig. 6-13b. The current flowing through the collector-side short circuit is, with good approximation, i_2' as $i_e = 0$ and $u_b/(r_c + r_{bb'})$ may be neglected. The

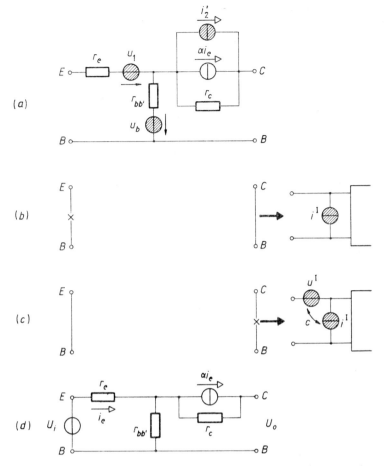

Figure 6-13 Calculation of transistor noise figure. *(a)* Modified equivalent circuit. *(b)* Terminations for the determination of i^I. *(c)* Terminations for the determination of u^I. *(d)* Calculation of the voltage gain.

output current reduced to the input [see Eq. (5-31)] is

$$i^I = \frac{i_2'}{\alpha} \tag{6-68}$$

Applying short circuit to the input and open circuit to the output (see Fig. 6-13c), the output voltage will be

$$u_0 = \alpha i_e r_c + i_2' r_c + i_e r_{bb'} + u_b \tag{6-69}$$

Only the first two terms are significant; the other terms may be neglected. The emitter current is

$$i_e = \frac{u_1 + u_b}{r_e + r_{bb'}} \tag{6-70}$$

From the equivalent circuit of Fig. 6-13d not comprising the noise generators, the open circuit voltage gain will be

$$A_U = \frac{U_{\text{out}}}{U_{\text{in}}} \cong \frac{\alpha r_c}{r_e + r_{bb'}} \tag{6-71}$$

Combining Eqs. (6-69) and (6-71), we have

$$u^{\text{I}} = \frac{u_0}{A} = u_1 + u_b + \frac{i_2'}{\alpha}(r_e + r_{bb'}) \tag{6-72}$$

The three terms on the right-hand side are independent of each other, so according to Eq. (5-20), we have

$$u^{\text{I}} u^{\text{I}*} = u_1^2 + u_b^2 + \frac{i_2' i_2'^*}{\alpha \alpha^*}(r_e + r_{bb'})^2 \tag{6-73}$$

Comparing Eqs. (6-68) and (6-72), it can be seen that there is a partial correlation between u^{I} and i^{I} as

$$u^{\text{I}*} i^{\text{I}} = \frac{i_2' i_2'^*}{\alpha \alpha^*}(r_e + r_{bb'}) = \frac{|i_2'|^2}{|\alpha|^2}(r_e + r_{bb'}) \neq 0 \tag{6-74}$$

The noise figure, which is influenced by the impedance of the generator connected to the input terminals, may be easily calculated with the aid of the two-generator equivalent circuit shown in Chapter 5. Let the resistive internal resistance of the signal source be R_g. Utilizing Eqs. (5-44) and (5-47), we then have

$$F = 1 + \frac{u^{\text{I}2} + i^{\text{I}2} R_g^2 + 2R_g \operatorname{Re}(u^{\text{I}*} i^{\text{I}})}{4kTR_g \Delta f}$$

$$= 1 + \frac{1}{4kTR_g \Delta f} \left\{ u_1^2 + u_b^2 + \left| \frac{i_2'}{\alpha} \right|^2 \left[(r_e + r_{bb'})^2 + R_g^2 + 2R_g(r_e + r_{bb'}) \right] \right\} \tag{6-75}$$

After substituting Eqs. (6-67), (6-37), and (6-51), we have

$$\boxed{F = 1 + \frac{r_e}{2R_g} + \frac{r_{bb'}}{R_g} + \frac{(r_e + r_{bb'} + R_g)^2}{2r_e R_g} \left(\frac{A - |\alpha|^2}{|\alpha|^2} + \frac{I_{CB0}}{I_E} \right)} \tag{6-76}$$

The same result was derived as early as 1957 by Nielsen.[102]

This formula has been derived utilizing the common base equivalent circuit. However, taking into account the neglections in Eq. (6-69), the input and output components of the one-port equivalent circuit comprising the signal source too (see Fig. 6-14a) may be separated, so the equivalent circuits shown in Figs. 6-14b and 6-14c are approximately equivalent.[3] This means that Eq. (6-76) and formulas derived from this equation are valid for the common emitter case too.

Following the approach utilized in Sec. 6-3, the separation of the different working point adjustments and frequency ranges will be again feasible in the following.

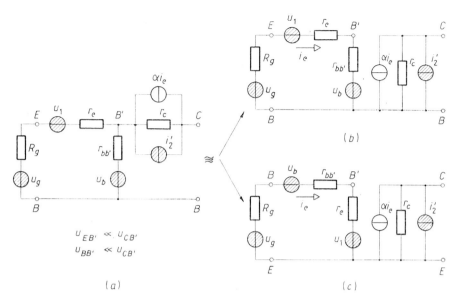

Figure 6-14 *(a)* Common base equivalent circuit. *(b)* and *(c)* Same kind of common base and common emitter equivalent circuits.

Low Frequency, Average Working Point

At low frequencies and average working point currents, $\alpha \cong \alpha_0 \cong A$. In this case,

$$\frac{A-|\alpha|^2}{|\alpha|^2} = \frac{1-\alpha}{\alpha} = \frac{1}{\beta} \tag{6-77}$$

and

$$F = 1 + \frac{r_e}{2R_g} + \frac{r_{bb'}}{R_g} + \frac{(r_e+r_{bb'}+R_g)^2}{2\alpha r_e R_g}\left(1-\alpha+\frac{\alpha I_{CB0}}{I_E}\right) \tag{6-78}$$

By applying a series of elementary rearrangements, Eq. (6-78) may also be utilized to calculate the noise figure of the common emitter equivalent circuit shown in Fig. 6-10. Thus

$$F = 1 + \frac{r_{bb'}}{R_g} + \frac{I_B}{2U_T}\frac{(R_g+r_{bb'})^2}{R_g} + \frac{I_C}{2U_T R_g \beta_0^2}\left[\frac{\beta_0 U_T}{I_C}+R_g+r_{bb'}\right]^2 \tag{6-79}$$

In contrast to Eq. (6-78), (6-79) is complicated because the currents I_B and I_C, characterizing the working point adjustment, are not eliminated. This may be an advantage if the flicker noise and burst noise have to be taken into account as their equivalent generators may simply be connected to the appropriate points of the base circuit. However, this equivalent circuit should not be applied at high frequencies because of the correlation shown by Eq. (6-60).

The base resistance $r_{bb'}$ and the current gain factor $\beta = \alpha/(1-\alpha)$ are production-dependent transistor parameters that cannot be altered later. On the other hand, the working point current and the signal source resistance may be chosen relatively freely. The noise figure shows a minimum value as a function of both of these parameters (see Fig. 6-15).

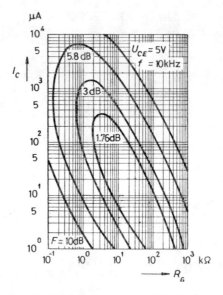

Figure 6-15 Noise figure curves of a Si planar transistor. (*Courtesy of Siemens.*)

Frequently, R_g is fixed. In this case, the working point emitter current pertaining to the minimum noise figure is derived by the derivation of Eq. (6-78):[162]

$$I_{E\,opt} = \frac{U_T}{\sqrt{1-\alpha(r_{bb'}+R_g)}} \tag{6-80}$$

and here

$$F_{min} = \left(1+\frac{r_{bb'}}{R_g}\right)\frac{1+\sqrt{1-\alpha}}{\alpha} \tag{6-81}$$

With a little rearrangement, Eq. (6-81) may written in the form

$$F_{min} = \left(1+\frac{r_{bb'}}{R_g}\right)\left(1+\frac{1}{\beta}+\frac{1}{\sqrt{\beta}}\right) \tag{6-82}$$

The two factors of this product show clearly the degradation relative to the ideal case of $F_{min}=1$. These factors may not be decreased arbitrarily as the increase in β causes an increase in $r_{bb'}$, and vice versa.

At optimum emitter current, we have from Eq. (6-79)

$$r_{e\,\text{opt}} = \sqrt{1-\alpha}\,(r_{bb'}+R_g) \ll r_{bb'}+R_g \cong R_g \tag{6-83}$$

which means that the last term in Eq. (6-78) may be simplified.[28] Thus

$$F \cong 1 + \frac{r_e/2+r_{bb'}}{R_g} + \frac{R_g}{2\beta r_e} \tag{6-84}$$

It is clearly seen that the noise figure has an extreme value as a function of R_g. Utilizing the calculation method applied in Sec. 5-2, simple noise equivalent circuits similar to those shown in Fig. 5-7 may be derived from Eq. (6-84) (see Fig. 6-16).

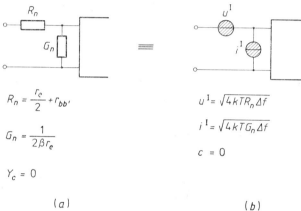

$$R_n = \frac{r_e}{2}+r_{bb'}$$

$$G_n = \frac{1}{2\beta r_e}$$

$$Y_c = 0$$

$$u^1 = \sqrt{4kTR_n\,\Delta f}$$

$$i^1 = \sqrt{4kTG_n\,\Delta f}$$

$$c = 0$$

(a)   (b)

Figure 6-16 Approximate equivalent circuits. *(a)* Comprising equivalent noise resistance and conductance. *(b)* Comprising voltage and current generators. (Uncorrelated noise sources: $Y_c=0$, $c=0$.)

Low Frequency, Low Current Working Point

At low frequencies and at low working point current, when the recombination taking place in the emitter–base junction may not be neglected, i.e. $A \neq \alpha_0$, we have

$$F = 1 + \frac{r_e}{2R_g} + \frac{r_{bb'}}{R_g} + \frac{(r_e+r_{bb'}+R_g)^2}{2r_e R_g}\left[A+\alpha_0^2\left(1-\frac{2A}{\alpha_0}\right)\right] \tag{6-85}$$

after substituting Eq. (6-48) into (6-78). A and α_0 are difficult to measure with sufficient accuracy for evaluating Eq. (6-85), so it is more appropriate to utilize the common emitter large-signal current-gain factor B and the small-signal current-gain factor β. After substituting Eqs. (6-63) and (6-64) and making elementary rearrangements, we arrive at the expression

$$F = 1 + \frac{r_e}{2R_g} + \frac{r_{bb'}}{R_g} + \frac{(r_e + r_{bb'} + R_g)^2}{2r_e R_g} \frac{1}{\beta} \frac{\beta + 1/m}{1 + \beta/m} \tag{6-86}$$

Compared with Eq. (6-78), the weight of the last term has increased m times.

High Frequency

At high frequencies, the frequency dependence of α is most significant. This is most easily taken into account using the form

$$\alpha = \frac{\alpha_0}{1 + jf/f_\alpha} \tag{6-87}$$

If $A \cong \alpha_0$, the fraction in Eq. (6-76) is easily rearranged to

$$\frac{A - |\alpha|^2}{|\alpha|^2} = \frac{\alpha_0}{|\alpha|^2} - 1 = \frac{1}{\alpha_0}\left[1 + \left(\frac{f}{f_\alpha}\right)^2\right] - 1 = \frac{1 - \alpha_0 + (f/f_\alpha)^2}{\alpha_0} \tag{6-88}$$

and so

$$F = 1 + \frac{r_e}{2R_g} + \frac{r_{bb'}}{R_g} + \frac{(r_e + r_{bb'} + R_g)^2}{2\alpha_0 r_e R_g}\left[(1 - \alpha_0) + \left(\frac{f}{f_\alpha}\right)^2\right] \tag{6-89}$$

The expression within the brackets is constant at low frequencies, and increases at the rate of 6 dB/octave at high frequencies. As in the Bode diagrams, the upper crossover frequency f_u may be introduced.

$$f_u = f_\alpha \sqrt{1 - \alpha_0} \cong \frac{f_\alpha}{\sqrt{\beta}} \tag{6-90}$$

because $\alpha_0 \cong 1$. It should be noted that this is only the crossover frequency of the last term of Eq. (6-89). To calculate the crossover frequency of the noise figure, the remaining terms have also to be taken into account.

If the signal source impedance is complex, $Z_g = R_g + jX_g$, and the numerator of the last term of Eq. (6-89) has to be changed to read $|r_e + r_{bb'} + Z_g|^2$.

Very High Frequency

The approximation (6-87) may not be used for $f > f_\alpha/2$. In addition, the frequency dependence of the emitter circuit generator source voltage u_1 in Fig. 6-13a has not been taken into account which is substantial in this frequency range, resulting in $G_e \neq 1/r_e$ in Eq. (6-67).

The calculations are rather complicated so for common base configuration only a few end results are presented here.[124]

$$|Y_g|_{opt} = \frac{1}{r_e}\sqrt{\frac{1/\beta+(f/f_\alpha)^2}{1+2\varrho+\varrho^2/\beta+\varrho^2(f/f_\alpha)^2}}$$ (6-91)

$$B_{g\,opt} = -\frac{1}{r_e}\frac{f/f_\alpha}{1+2\varrho+\varrho^2/\beta+\varrho^2(f/f_\alpha)^2}$$ (6-92)

$$F_{min} = 1+\varrho\left(\frac{f}{f_\alpha}\right)^2\left[1+\sqrt{1+\frac{2}{\varrho(f/f_\alpha)^2}}\right] \quad \text{if} \quad F_{min} > 4.8\ \text{dB}$$ (6-93)

where $\varrho = r_{bb'}/r_e$. The shape of the function $F_{min}(f)$ and characteristic data of its tangents are shown in Fig. 6-17. In case of imperfect input matching, the noise figure will change according to Eq. (5-61) in the vicinity of F_{min}.

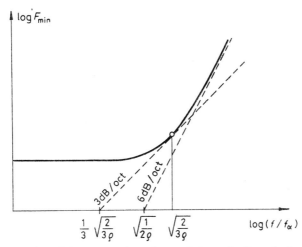

Figure 6-17 Frequency dependence of the noise figure in the high-frequency range according to Eq. (6-93).

Recently, efforts have been made to make the Nielsen formula [Eq. (6-76)] more accurate. There are solutions for complex $Z_e \neq r_e$,[158] taking into account the emitter–base space charge capacitance.[47] The latter is of importance because the upper cut-off frequency of transistors having inhomogeneous base doping is limited practically not by the base transit time but by the emitter–base time constant. Taking this into account,

$$F = 1+\frac{r_{bb'}}{R_g}+\frac{r_e}{2R_g}+\left[\left(1+\frac{f^2}{f_\alpha^2}\right)\left(1+\frac{f^2}{f_e^2}\right)-\alpha_0\right]\frac{(R_g+r_{bb'}+r_e)^2}{2\alpha_0 r_e R_g}$$ (6-94)

where

$$f_e = \frac{R_g+r_{bb'}+r_e}{2\pi C_{Te} r_e (R_g+r_{bb'})}$$ (6-95)

Rather complicated formulas apply to the equivalent noise generators and noise figure of the common emitter configuration.[26]

The lead reactances are responsible for further difficulties (see Fig. 6-8). These influence primarily the value of the optimum source admittance and, after matching has been provided, the expression (6-93) may be used to calculate F_{min} even in these cases. Alternatively, Eq. (6-94) may be utilized.

Flicker Noise

The main source of flicker noise is the emitter–base junction and so can be taken into account by an additional current generator connected between points E and B' in the T or π equivalent circuits treated earlier.

Let us utilize the experimentally determined relation

$$i_b^2 = KI_B^\gamma \left(\frac{f_c}{f}\right) \Delta f \tag{6-96}$$

This relation comprises, in a generalized form, the considerations given in connection with Eqs. (6-62) to (6-65). From (6-79) we then have

$$F = 1 + \frac{r_{bb'}}{R_g} + \frac{1}{2}\frac{(R_g+r_{bb'})^2}{R_g}\left[\frac{I_B}{U_T} + \frac{KI_B^\gamma}{4kT}\left(\frac{f_c}{f}\right)\right] + \frac{I_C}{2U_T R_g \beta_0^2}\left(\frac{\beta_0 U_T}{I_C}+R_g+r_{bb'}\right)^2 \tag{6-97}$$

This function has a flat minimum for $R_g = r_{bb'}$, and here

$$F_{min} = 2 + \frac{U_T}{2I_C r_{bb'}} + \frac{r_{bb'}}{kT} KI_B^\gamma\left(\frac{f_c}{f}\right) \tag{6-98}$$

The signal source resistance needed for this minimum noise figure, equal to $r_{bb'}$, is generally much less than the value applying to the white-noise range. The value calculated by the derivation of Eq. (6-84) with respect to R_g is

$$R_{g\,opt} = \sqrt{\beta r_e^2 + 2\beta r_e r_{bb'}} \tag{6-99}$$

The situation becomes more complicated if two flicker-noise sources, similarly to Fig. 6-12, have to be taken into account [now one of them i_{f1} being connected to a part $xr_{bb'}$ $(x<1)$ of the base resistance and the other i_{f2} is connected to the points $B'–E$]. In this case, the noise figure increment due to the flicker noise[60] is

$$\Delta F = \frac{i_{f1}^2(R_g+xr_{bb'})^2 + i_{f2}^2(R_g+r_{bb'})^2}{4kTR_g \Delta f} \tag{6-100}$$

which has a minimum at the source resistance

$$R_{g\,opt} = r_{bb'}\sqrt{\frac{x^2+\varepsilon^2}{1+\varepsilon^2}} \qquad \varepsilon^2 = \frac{i_{f2}^2}{i_{f1}^2} \tag{6-101}$$

In the transition region between the flicker noise and the white noise [see Eq. (6-97)], the noise figure has been approximated by simple addition. However, this requires that the added terms should be independent, which is not necessarily true. For instance, if the flicker noise is generated through thermal feedback,[96] it is correlated with the originating cause, i.e. the white noise.

The $1/f$-type frequency response is frequently upset by the burst noise.[163] The power spectrum of the two-state burst noise, according to Eq. (4-109), is

$$S(f) = \vartheta \frac{\text{const.}}{(v_{19} + v_{29})^2 + \omega^2}$$

This may be superimposed on the $1/f$ spectrum at any point (see Fig. 6-18). A more accurate analysis shows that the terms originating from the cross correlation of the flicker noise and pulse noise may generally be neglected.[90]

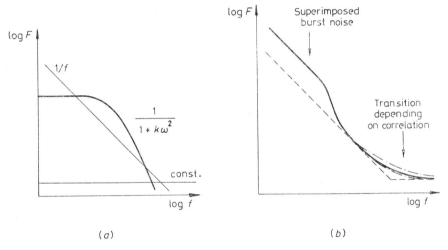

Figure 6-18 *(a)* Frequency response of white noise, flicker noise, and burst noise. *(b)* Overall noise figure.

6-5 TRANSISTORS WITHIN INTEGRATED CIRCUITS

The internal equivalent circuit of transistors within integrated circuits with solid state substrates does not differ from that shown in Fig. 6-6; between the collector and the collector lead, there is sometimes a series resistance which may not be neglected. Furthermore, it has to be taken into account that the collector of an integrated transistor is embedded in a solid state substrate, separated from the transistor only by a reverse biased p–n junction. Figure 6-19a shows the structure of an integrated circuit transistor, together with additional elements that are significant from the noise aspect.[127]

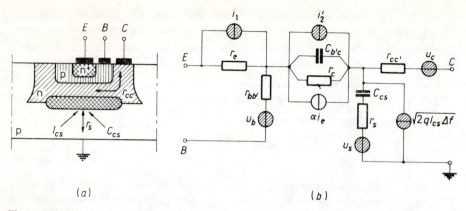

Figure 6-19 Integrated circuit transistor. *(a)* Cross sectional view. *(b)* Noise equivalent circuit

The solid state substrate has a finite resistance r_s, generating a noise voltage $u_s = \sqrt{4kTr_s\Delta f}$. This noise voltage reaches the collector through the collector–substrate capacitance C_{CS}. The shot noise in the reverse current flowing through the collector–substrate p–n junction is also observed in the collector circuit. Furthermore, the thermal-noise voltage $u_c = \sqrt{4kTr_{cc'}\Delta f}$ of the series collector resistance $r_{cc'}$ has also to be taken into account. These parameters are included in the noise equivalent circuit shown in Fig. 6-19b.

At low frequencies, C_{CS} represents an open circuit, and

$$r_{cc'} \ll \left| r_c \middle\| \frac{1}{j\omega C_{b'c}} \right| \quad \text{and} \quad u_c \ll \left| \alpha i_e \left(r_c \middle\| \frac{1}{j\omega C_{b'c}} \right) \right|$$

This means that only I_{CS} is effective, so there is a single additional term in Eq. (6-78). Thus

$$F = 1 + \frac{r_e}{2R_g} + \frac{r_{bb'}}{R_g} + \frac{(r_e + r_{bb'} + R_g)^2}{2\alpha r_e R_g}\left[1 - \alpha + \frac{I_{CS}}{I_C}\right] \tag{6-102}$$

The new collector circuit elements shown in Fig. 6-19b would only be effective at high frequencies at which the operation of conventional analog integrated circuits is no longer possible due to other reasons (e.g., Miller capacitance). If it is still necessary to take these effects into account, reference is made to calculations published in the literature.[127]

The previous expressions may be applied, with due considerations, for the determination of the low-frequency noise sources and noise figure of the p–n–p lateral transistor. However, it should be borne in mind that the current-gain factor of these transistors is generally low (the base current is high) so the current partition noise is much more significant than in high-gain p–n–p transistors. It would be difficult to present reliable formulas for high-frequency noise relations. The high-frequency small-signal operation has not yet been explained completely.[109, 122]

Equation (6-97), and possibly (6-100), may be applied to all integrated circuit transistors in the flicker region. The condition for optimization of these formulas is $R_g = r_{bb'}$. For low R_g, the advantages due to integration may be utilized: the resultant $r_{bb'}$ may be decreased by parallel connection of several transistors (integration), thus realizing the optimum matching condition.[171]

The imperfect structure may give rise to burst noise, both in discrete and integrated transistors. In the early stage of operational amplifier development, the reject level was high as a single transistor having high burst noise could spoil the complete integrated circuit. On the other hand, it is important to note that the noise sources, even in adjacent transistors, are uncorrelated for all kind of noise sources (white noise, $1/f$ noise, and burst noise).[13, 20, 56]

Special attention should be given to the differential transistor pair supplemented by the emitter current source (see Fig. 6-20). This arrangement is widely used both with discrete transistors and as the first stage of integrated operational amplifiers. Figure 6-20*b* shows the two-generator noise equivalent circuit reduced to the transistor inputs.

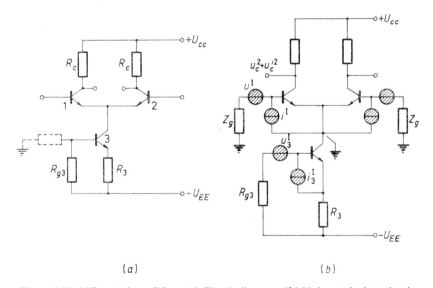

(a)   (b)

Figure 6-20 Differential amplifier. *(a)* Circuit diagram. *(b)* Noise equivalent circuit.

The connected emitter electrodes of the differential pair may be regarded to be at AC ground potential. According to Eq. (5-46), the squared value of the noise voltage generated at each transistor base is

$$u_b^2 = u_g^2 + u^{I2} + |Z_g i^{I}|^2 + 2\,\mathrm{Re}\,(u^{I*} Z_g i^{I}) \tag{6-103}$$

Taking into account the transistor voltage gain $g_m R_c$, the squared noise voltage at the collector is

$$u_c^2 = g_m^2 R_c^2 u_b^2 \qquad (6\text{-}104)$$

To this has to be added the component originating from the squared fluctuation $i_{ee}^2/2$ of the working point current $I_{EE}/2$, which is $i_{ee}^2 \alpha^2 R_c^2/2 = u_c'^2$.

The useful output signal of the differential amplifier is normally the voltage difference between the two collectors. Onto this difference is superimposed twice the value of the squared noise voltage calculated according to Eq. (6-103). The squared sum is explained by the fact that the noise sources of the first and second transistors are independent of each other. It seems that, even with optimum input matching, the differential amplifier is at least twice as noisy as the asymmetrical amplifier of similar structure. However, the internal resistances Z_g of the two signal sources are actually connected in series (see Fig. 6-20), so the input voltage pertaining to the same input signal power is increased $\sqrt{2}$ times. This means that the noise figure remains unchanged. In case of perfect symmetry, u_c' appears at the two collectors with the same phase and amplitude, so it has no effect on the differential output signal.

A separate analysis is justified for the noise of the current-controlled current generator shown in the lower part of Fig. 6-20a, as this configuration is frequently applied to integrated circuits (e.g., phase addition). Let us assume that the noise equivalent circuits of both the diode-connected and the normally connected transistor correspond to Fig. 6-10, with the extension that the noise current generator i_{bf} also includes the flicker components. This generator is connected between the emitter and a part of base resistance characterized by $xr_{bb'}$, where $x < 1$ is a relative number originating from the base lead. The white-noise component of the output current is then given by[12]

$$i_{ow} = \sqrt{2\,2qI_C \Delta f}\; \frac{[1 + 2b + b^2/\beta + 2d(1 + b/\beta) + 2d^2/\beta]^{1/2}}{1 + b/\beta + d} \qquad (6\text{-}105)$$

and the flicker component is

$$i_{of} = \sqrt{2}\,i_{bf}\; \frac{[1 + (xb)^2 + 2d(1 + xb) + 2d^2]^{1/2}}{1 + b/\beta + d} \qquad (6\text{-}106)$$

where $b = r_{bb'}/r_e$ and $d = R_E/r_e$.

This circuit appears at an important point of operational amplifiers comprising current mirror input stages (see Fig. 6-21).[41] After applying simplifying assumptions, it can be shown that all noise sources may be concentrated into a single current generator with a source parameter

$$i_n^2 = 4qI^+ \Delta f\left(1 + \frac{r_{bb'}\, qI^+}{kT}\right) \qquad (6\text{-}107)$$

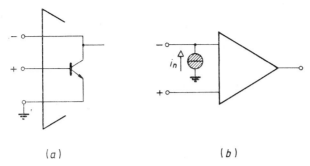

(a)   (b)

Figure 6-21 *(a)* Operational amplifier with current mirror
input. *(b)* Illustration of Eq. (6-107), I^+ is the bias
current of noninverting input

6-6 CHANNEL NOISE OF FIELD-EFFECT TRANSISTORS[142]

The schematic structures of two of the most frequently used field-effect transistors
are shown in Fig. 6-22; only the significant parts for operation (i.e. the channel
and the gate electrode) are shown. Figures 6-22*a* and 6-22*b* show the structures
of field-effect transistors with p–n junction and isolated gate electrode, respectively. In spite of their similar geometrical layouts, there are fundamental differences in their physical operation. In a p–n junction device, the current flows
inside the semiconductor material, while in a MIS (Metal-Insulator-Semiconduc-
tor) or MOS (Metal-Oxide-Semiconductor) the current flow is confined to an
extremely thin layer on the surface. These differences appear naturally in the noise
parameters too.

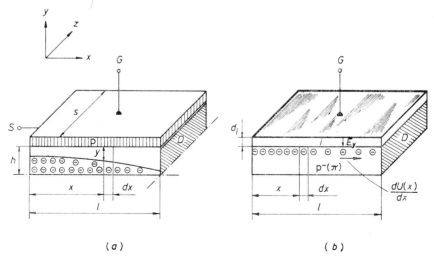

(a)   (b)

Figure 6-22 Simplified model of field-effect transistor.
(a) p–n junction device. *(b)* MIS device, I^+ is the bias current of noninverting input.

For a unified treatment, let us suppose that both devices be N-type, i.e. the charge carriers are electrons. Then, in the normal range of operation, $U_{DS}>0$. The device shown in Fig. 6-22a operates in the depletion mode, that in Fig. 6-22b operates in the enhancement mode. Therefore, the realization of the first condition requires a negative control voltage $U_{GS}\leq0$, while the realization of the second condition requires that $U_{GS}\geq0$.

As $I_G\ll I_D$, I_D is constant along the whole channel and, according to definition,

$$I_D = \frac{dQ}{dt} \tag{6-108}$$

The charge in a section of length dx of the channel of the device shown in Fig. 6-22a is

$$dQ = qnA\,dx = qns(2y)\,dx \tag{6-109}$$

where n is the electron density in unit volume. By substituting $2y$, the symmetry of the device has been taken into account.

According to the Gauss theorem, the surface charge density on the semiconductor bulk surface in the device shown in Fig. 6-22b due to the voltage given on the control electrode is

$$Q_S = D = \varepsilon_0\varepsilon_i E_y(x) \tag{6-110}$$

where $\varepsilon_0 = 8.85\times10^{-12}$ F/m and ε_i is the relative dielectric constant of the insulator. The quantity applying to the band sdx is

$$dQ = Q_S s\,dx \tag{6-111}$$

As $dx/dt = v_x = \mu|E_x| = \mu dU_x/dx$, substituting Eqs (6-109) and (6-111) into (6-108), we have

$$I_D = qns(2y)\mu\frac{dU_x}{dx} \tag{6-112a}$$

$$I_D = \varepsilon_0\varepsilon_i E_y s\mu\frac{dU_x}{dx} \tag{6-112b}$$

Taking into account that $(2y)$ and E_y are functions of U_x—the potential along the channel—the differential equations (6-112a) and (6-112b) are separable and may be solved in the general form

$$\int_0^l I_D\,dx = \text{const.} \int_0^{U_{DS}} f(U_x)\,dU_x \tag{6-113}$$

The channel height of the p–n junction device at point x is

$$y = h\left[1-\left(\frac{U_x+|U_{GS}|}{U_P}\right)^m\right] = \frac{f_1(U_x)}{2} \tag{6-114}$$

where U_P is the voltage required for the complete pinch-off of the channel and m is an exponent that depends on the dopant profile. On the other hand, the field strength in the insulating layer of width d_i in a MIS transistor is

$$E_y = \frac{U_{GS} - V_T - U_x}{d_i} = f_2(U_x) \qquad (6\text{-}115)$$

where $V_T = 0$ if all lines of field starting from the control electrode terminate in movable electrons. However, $V_T \neq 0$ and may be either positive or negative. This is explained partly by the substrate dopants and partly by the unneutralized charges within the insulator (oxide) and on the insulator–semiconductor interface, respectively.

Substituting Eq. (6-114) into (6-113), the equation of the p–n junction field-effect transistor characteristic is

$$I_D = G_0 \left\{ (|U_{GS}| + U_{DS}) \left[1 - \frac{1}{m+1} \left(\frac{|U_{GS}| + U_{DS}}{U_P} \right)^m \right] \right.$$

$$\left. - |U_{GS}| \left[1 - \frac{1}{m+1} \left(\frac{|U_{GS}|}{U_P} \right)^m \right] \right\} \qquad (6\text{-}116)$$

where

$$G_0 = \frac{2hsq\mu n}{l} \quad \frac{\text{A}}{\text{V}} \qquad (6\text{-}117)$$

Similarly for the MIS transistor, we have

$$I_D = \gamma \left[(U_{GS} - V_T) U_{DS} - \frac{U_{DS}^2}{2} \right] \qquad (6\text{-}118)$$

where

$$\gamma = \frac{\varepsilon_0 \varepsilon_i \mu s}{l d_i} \quad \frac{\text{A}}{\text{V}^2} \qquad (6\text{-}119)$$

Equations (6-116) and (6-118) apply only in the triode range for which $I_D = f(U_{GS}, U_{DS})$; the former is applicable for $U_{DS} + |U_{GS}| \leq U_P$, and the latter for $U_{GS} - U_{DS} \geq V_T$.

If U_{DS} exceeds the above limits, i.e. if the channel is pinched off at the drain, then

$$I_{DS} = G_0 \left\{ \frac{m}{m+1} U_P - |U_{GS}| \left[1 - \frac{1}{m+1} \left(\frac{|U_{GS}|}{U_P} \right)^m \right] \right\} \qquad (6\text{-}120)$$

will apply instead of (6-116), and

$$I_{DS} = \frac{\gamma}{2} (U_{GS} - V_T)^2 \qquad (6\text{-}121)$$

will apply instead of (6-118). In the pinch-off region, the mutual conductance is

$$g_{ms} = \frac{dI_{DS}}{dU_{GS}} = G_0\left[1 - \left(\frac{|U_{GS}|}{U_P}\right)^m\right] \qquad (6\text{-}122)$$

and

$$g_{ms} = \gamma(U_{GS} - V_T) \qquad (6\text{-}123)$$

We shall in addition need the characteristic value at the origin, $g_c = \dfrac{dI_{DS}}{dU_{GS}}\Big|_{U_{GS}=0}$.
It is easily shown that in both cases this is equal to g_{ms}.

Utilizing the similarity between the mathematical models of the two devices, only the MIS transistor noise sources will be analyzed in detail.[170] However, our results can easily be extended to the other device.

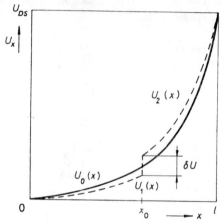

Figure 6-23 MIS transistor channel potential
as a function of distance. (——) equilibrium
state, (- - -) perturbed state.

Figure 6-23 shows the potential distribution $U(x)$ of the channel. Let us suppose that for some reason, a potential perturbation δU appears at point x_0. This will change the potential distribution both in the range $0 < x < x_0$ and $x_0 < < x < 1$, and consequently, the channel current. The change in the channel current may be calculated by dividing the transistor into two parts ($x < x_0$ and $x > x_0$), and for both parts, the current I_{D0} pertaining to curve $U_0(x)$ and the currents I_{D1} and I_{D2} pertaining to curves $U_1(x)$ and $U_2(x)$ are calculated. Utilizing Eqs. (6-112) and (6-115), the current differences are

$$I_{D1} - I_{D0} = \gamma I\left\{[U_{GS} - V_T - U_1(x)]\frac{dU_1}{dx} - [U_{GS} - V_T - U_0(x)]\frac{dU_0}{dx}\right\} \qquad (6\text{-}124)$$

and

$$I_{D2} - I_{D0} = \gamma I\left\{[U_{GS} - V_T - U_2(x)]\frac{dU_2}{dx} - [U_{GS} - V_T - U_0(x)]\frac{dU_0}{dx}\right\} \qquad (6\text{-}125)$$

Equation (6-124) applies in the range $0<x<x_0$, so after separating the variables and integrating according to Eq. (6-113) we get $(I_{D1}-I_{D0})x_0$ on the left-hand side, and on the right $U_1(x_0)$, $U_0(x_0)$, and $U_1(0)=U_0(0)=0$ are to be substituted. In a similar way, the left-hand side of the Eq. (6-125) is calculated to be $(I_{D2}-I_{D0})(l-x_0)$, with $U_2(x)$ and $U_0(x)$ to be taken into account at the limits x_0 and l. Due to the current continuity,

$$I_{D1}-I_{D0} = I_{D2}-I_{D0} = \delta I \qquad (6\text{-}126)$$

will be applicable to the current difference between the perturbed and unperturbed channels. So the addition of the solutions of differential equations (6-124) and (6-125), utilizing the expressions

$$\delta U = U_2(x_0)-U_1(x_0)$$

and

$$U_0(x_0) = \frac{U_2(x_0)+U_1(x_0)}{2}$$

will result in the expression

$$\delta I = -\gamma [U_{GS}-V_T-U_0(x_0)]\delta U \qquad (6\text{-}127)$$

The potential perturbation δU may be, for instance, the consequence of the thermal noise in the channel element of resistance dr. Thus

$$\overline{\delta U^2} = 4kT\,dr\Delta f = \frac{4kT\,\Delta f}{I_{D0}}\,dU \qquad (6\text{-}128)$$

as the relation $dU=I_{D0}dr$ is valid at any place, due to the distance-independent channel current. The overall (short circuit) noise current of the field-effect transistor is determined by quadratically adding the current fluctuations generated by independent perturbations at different channel points. Thus

$$i_d^2 = \int_0^{U_{DS}} \overline{(\delta I)^2} = 4kT\Delta f\,\frac{\gamma^2}{I_D}\int_0^{U_{DS}} [U_{GS}-V_T-U(x_0)]^2\,dU \qquad (6\text{-}129)$$

Substituting γ/I_D from Eq. (6-118) and introducing the notation

$$\eta = U_{DS}/(U_{GS}-V_T),$$

we have

$$i_d^2 = 4kT\Delta f\gamma (U_{GS}-V_T)\frac{1-\eta+\eta^2/3}{1-\eta/2} \qquad (6\text{-}130)$$

After substituting Eq. (6-123), the final result is

$$\boxed{i_d^2 = 4kT\Delta f g_{ms} K_d(\eta)} \qquad (6\text{-}131)$$

where $1>K_d(\eta)>2/3$ if $0<\eta<1$. This function, applicable to all MIS devices as a first approximation, is shown by Fig. 6-24a.

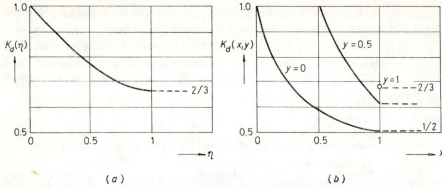

Figure 6-24 Channel noise working point dependence in a field-effect transistor. *(a)* MIS device. *(b)* p–n junction device with abrupt dopant profile.

For $U_{DS}=0, \eta=0$ and $K_d(\eta)=1$. The result for this case is similar to Eq. (4-51), according to which the noise-current spectrum density of conductance g_{ms} is given by $S_i(f)=4kTg_{ms}$. But $g_{ms}=g_c(U_{DS}=0)$, i.e. in this case, the noise behavior of the field-effect transistor is exactly the same as that of a linear ohmic resistance in thermal equilibrium, as foreseen. The derivation Equations (6-124) to (6-131) may be performed also for a p–n junction device, and results will depend on the dopant profile. For an abrupt dopant profile, i.e. $m=1/2$, Eq. (6-131) will be valid with the substitution of $K_d(x, y)$,[142] where

$$K_d(x, y) = \frac{x-y-\dfrac{4}{3}(x^{3/2}-y^{3/2})+\dfrac{1}{2}(x^2-y^2)}{(1-y^{1/2})[(x-y)-\dfrac{2}{3}(x^{3/2}-y^{3/2})]} \qquad (6\text{-}132)$$

and

$$x = \frac{|U_{GS}|+U_{DS}}{U_P} \qquad y = \frac{|U_{GS}|}{U_P} \qquad (6\text{-}133)$$

For $U_{DS}=0$, $x=y$, and $K_d(x, y)=1$, which is the same as for an MIS device. Figure 6-24b shows $K_d(x)$ with the parameters $y=0$, 1/2, and 1. The curves may not be interpreted for $x<y$.

Equations (6-130) and (6-132) are, strictly speaking, valid only for the triode mode of operation. However, in the absence of a better approximation, their use may be extended to the pinch-off region, which is more important for practical purposes (see the dashed lines in Fig. 6-24).

In more rigorous treatments, the field-effect transistor is regarded as a distributed parameter active network.[23, 71, 105, 123, 156, 160] The results thus obtained do not significantly differ from those obtained above. During the above considerations, it has been implicitly assumed that the pinch-off part of the channel is negligibly short compared with the total length, therefore the former has no substantial effect on the characteristics and noise properties of the device. However, technological improvements, primarily in order to achieve higher cut-off frequencies, have resulted in increasingly shorter channels having the following properties:

1. The average field strength has increased.
2. As a consequence, the pinch-off part of the channel has become relatively longer.
3. The relation $v = \mu E$ utilized for Eq. (6-112) is no longer valid, and the electrons move with saturated velocity.
4. The energy of the high-velocity electrons is higher than that calculated from the temperature of lattice, so these electrons may be regarded as "hot" electrons.

At the time of the revision of this book, hot electron effects have been intensively investigated,[177,178] but no unified theory has yet been evolved. However, several valuable papers have been published presenting the consequences of the effects listed above in silicon, and primarily, in gallium arsenide field-effect transistors.[8,59,107,126,129,131]

Naturally, the actual field-effect transistor does not comprise only a single ideal channel. The resistances $r_{ss'}$ and $r_{dd'}$ of leads S and D, not shown in Fig. 6-22, also have thermal noises. The noise equivalent circuit of the channel thus extended is shown in Fig. 6-25.

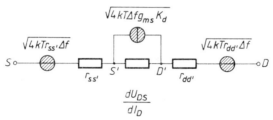

Figure 6-25 Noise equivalent circuit of channel in field-effect transistor comprising lead elements.

6-7 GATE NOISE OF FIELD-EFFECT TRANSISTORS

A field-effect transistor with a p–n junction always has a gate current I_G. This may be made up of three components (if the minority carrier density of the highly doped gate may be neglected):

1. The minority carriers thermally generated in the channel.
2. The carrier pairs thermally generated in the depletion layer.
3. The minority carriers generated in the channel by impact ionization.

In silicon devices, normally the second component dominates, but at high temperatures, the first component may also be observed. The third component is observed if the majority carriers are sufficiently accelerated by the high field strength in the channel for the generation of new carrier pairs. As the probability of ionization is higher for electrons than for holes, the ionization component is observable earlier in the gate electrode current of n-channel devices.[2,115]

In order to calculate the noise current, the components have to be multiplied by following factors: first component—unity, second component—approximately 2/3 according to Eq. (6-30), third component—approximately unity.[114] Thus

$$\overline{i_g^2} = 2q\Delta f\left(I_0 + \frac{2}{3}I_{Gen} + I_{Ion}\right) = 2qgI_G\,\Delta f \qquad (6\text{-}134)$$

where I_G is the complete gate current and g is a weighting factor depending on the ratio of the components.

In both basic types of field-effect transistors, a phenomenon similar to the induced grid noise of electron tubes may be observed.[139] This is because the voltage fluctuations generated in the channel reach the gate electrode through the channel–gate capacitance. The detailed calculation will again be performed only for the MIS device.[170]

According to Eqs. (6-110) and (6-115), the surface charge densities pertaining to the equilibrium and perturbed potential distribution according to Fig. 6-23 are

$$Q_{S0} = \frac{\varepsilon_0\varepsilon_i}{d_i}[U_{GS} - V_T - U_0(x)]$$

$$Q_{S1} = \frac{\varepsilon_0\varepsilon_i}{d_i}[U_{GS} - V_T - U_1(x)] \qquad (6\text{-}135)$$

$$Q_{S2} = \frac{\varepsilon_0\varepsilon_i}{d_i}[U_{GS} - V_T - U_2(x)]$$

To calculate the whole charge perturbation, the surface charge density differences $Q_{S1} - Q_{S0}$ and $Q_{S2} - Q_{S0}$—which are position-dependent—have to be integrated for the complete surface of the gate electrode. Thus

$$\delta Q = \frac{\varepsilon_0\varepsilon_i S}{d_i}\left[\int_0^l U_0(x)\,dx - \int_0^{x_0} U_1(x)\,dx - \int_{x_0}^l U_2(x)\,dx\right] \qquad (6\text{-}136)$$

Combining Eqs. (6-112), (6-115), and (6-119), we have

$$\frac{dx}{dU_x} = \frac{\gamma l}{I_D}[U_{GS} - V_T - U_x] \qquad (6\text{-}137)$$

which may be utilized to rewrite the integration variables of Eq. (6-136). Thus

$$\delta Q = \frac{\varepsilon_0 \varepsilon_i s}{d_i} \gamma l \left\{ \int_0^{U_{PS}} \frac{[U_{GS} - V_T - U_0(x)] U_0(x)}{I_{D0}} dU_0 \right.$$

$$- \int_0^{U(x_0)} \frac{[U_{GS} - V_T - U_1(x)] U_1(x)}{I_{D0} + \delta I} dU_1 - \int_{U(x_0)}^{U_{PS}} \frac{[U_{GS} - V_T - U_2(x)] U_2(x)}{I_{D0} + \delta I} dU_2 \right\}$$

$$(6\text{-}138)$$

where I_{D0} may be interpreted by Eqs. (6-124) and (6-125), and δI may be interpreted by Eq. (6-126). After calculating the integral and neglecting the second-order small terms, the surface charge difference is

$$\delta Q = \frac{\delta I}{I_{D0}} \left[Q_0 - \frac{\varepsilon_0 \varepsilon_i s l}{d_i} U(x_0) \right] \tag{6-139}$$

where

$$Q_0 = \frac{\varepsilon_0 \varepsilon_i s}{d_i} \frac{\gamma l}{I_{D0}} \left[(U_{GS} - V_T) \frac{U_{DS}^2}{2} - \frac{U_{DS}^3}{3} \right] \tag{6-140}$$

By substituting Eqs. (6-127) and (6-128), the mean square value of the gate charge fluctuation is

$$q_g^2 = \int_0^{U_{PS}} M(\delta Q)^2 \, dU$$

$$= \left(\frac{\varepsilon_0 \varepsilon_i s l}{d_i} \right)^2 4kT \Delta f \frac{\gamma^2}{I_{D0}^3} \int_0^{U_{PS}} [U_{GS} - V_T - U(x)] \left[\frac{Q_0 d_i}{\varepsilon_0 \varepsilon_i s l} - U(x) \right] dU$$

$$= 4kT \Delta f \frac{C_{gc}^2}{g_{ms}} K_g(\eta) \tag{6-141}$$

Here

$$C_{gc} = \frac{\varepsilon_0 \varepsilon_i s l}{d_i} \tag{6-142}$$

is the gate–channel capacity and K_g is a complicated function of $\eta = U_{DS}/(U_{GS} - V_T)$ (see Fig. 6-26a). For $0 < \eta < 1$, $1/12 < K_g(\eta) < 16/135 \cong 0.12$. Finally, the squared fluctuation of the gate current is

$$\boxed{i_g^2 = \omega^2 q_g^2 = 4kT \Delta f \frac{\omega^2 C_{gc}^2}{g_{ms}} K_g(\eta)} \tag{6-143}$$

Both the channel noise investigated in Sec. 6-6 and the induced gate noise treated here are brought about by random fluctuation of the channel potential,

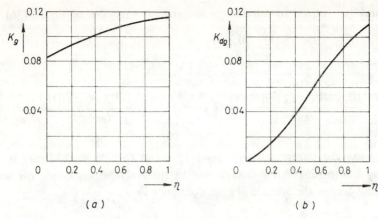

Figure 6-26 Working point dependence of MIS transistor noise.
(a) Gate electrode noise. *(b)* Cross correlation noise.

so there is a correlation between i_d and i_g. To calculate this, let us multiply Eq. (6-139) by δI and substitute Eqs. (6-127) and (6-128). Thus

$$\delta Q \delta I = 4kT\Delta f \frac{\gamma^2}{I_{D0}^2} [U_{GS}-V_T-U(x_0)]^2 \left[Q_0 - \frac{\varepsilon_0 \varepsilon_i sl}{d_i} U(x_0)\right] dU \quad (6\text{-}144)$$

After integration, we have

$$i_g^* i_d = j\omega q_g^* i_d = j\omega \int_0^{U_{DS}} (\delta Q \delta I)\, dU = 4kT\Delta f j\omega C_{gc} K_{dg}(\eta) \quad (6\text{-}145)$$

if $0<\eta<1$, $0<K_{dg}(\eta)<1/9$ (see Fig. 6-26b).

For practical purposes, again the pinch-off region, for which the value pertaining to $\eta=1$ is extrapolated, is of importance. Then Eqs. (6-131), (6-143), and (6-145) will take the forms

$$\boxed{i_d^2 = i_d i_d^* = \frac{2}{3}\, 4kT\, g_{ms}\, \Delta f} \quad (6\text{-}146)$$

$$\boxed{i_g^2 = i_g i_g^* = \frac{16}{135}\, 4kT\, \frac{\omega^2 C_{gc}^2}{g_{ms}}\, \Delta f} \quad (6\text{-}147)$$

$$\boxed{i_g^* i_d = \frac{1}{9}\, 4kT j\omega C_{gc}\, \Delta f} \quad (6\text{-}148)$$

Utilizing the definition (5-32), the complex correlation factor is

$$c = \frac{i_g^* i_d}{\sqrt{|i_g|^2 |i_d|^2}} = 0.39j \quad (6\text{-}149)$$

This factor is independent of the frequency, the mutual conductance, and the capacity C_{gc}. The physical reason for this is the purely capacitive coupling between the two generators.

Figure 6-27 shows the two-generator noise equivalent circuits of the field-effect transistor. K_d, K_g, and K_{dg} are independent of the working point.

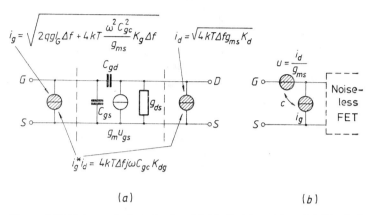

(a)   (b)

Figure 6-27 Noise equivalent circuits of field-effect transistor. *(a)* Physical equivalent circuit. *(b)* Noise sources reduced to the input.

The above coefficients have different values for the junction field-effect transistor.[44, 66, 143] Figure 6-28 shows the values for K_g and K_{dg} at the limit of pinch-off as a function of the variable $y=|U_{GS}|/U_P$.

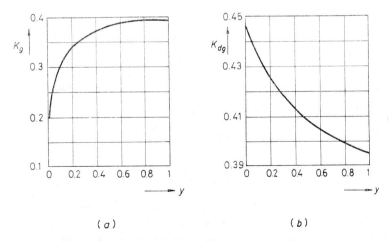

(a)   (b)

Figure 6-28 Working point dependence of junction field-effect transistor noise. *(a)* Gate electrode noise. *(b)* Cross correlation noise.

In the literature, there are more detailed equivalent circuits than that shown in Fig. 6-27 and also several versions of noise figure calculations.[4, 67, 83, 108, 113, 145] Calculations taking into account the hot electron effect are especially lengthy.[107] Furthermore, it is not generally true that the channel noise and the gate electrode noise are additive.

6-8 OTHER NOISE SOURCES IN FIELD-EFFECT TRANSISTORS

In contrast to MIS and bipolar transistors whose operation is necessarily affected by surface effects, the field-effect transistor having a p–n junction is basically a volume device. Consequently, such a transistor is almost free of side effects (e.g., flicker noise) which are difficult to calculate. At medium and high frequencies, the noise parameters that can be measured are well approximated by the theoretical values calculated in Secs. 6-6 and 6-7.[15, 75, 86, 87] In some cases, the lead resistances have to be taken into account, according to the equivalent circuit shown in Fig. 6-25. At high frequencies, the capacitances C_{gs} and C_{gd} between the gate electrode and the other two electrodes may be significant.

If there are traps in the depletion layer below the gate electrode because of imperfections in the semiconductor crystal (region $h–y$ in Fig. 6-22a) then generation–recombination noise may be generated.[31, 42, 43, 81, 116] This is explained by the fact that the width of the depletion layer is modulated by the occupied or nonoccupied state of the traps. If the generation–recombination process is characterized by a continuous spectrum of time constants instead of a single time constant, the characteristic $1/f$ spectrum may prevail in a certain region. In spite of this, of all the semiconductor devices used presently, the low-frequency excess noise is least in junction field-effect transistors,[65, 77] and the order of magnitude of f_c defined by Eq. (6-96) may even be of the order of 10 Hz. In the equivalent circuit of Fig. 6-29b, the flicker-noise source may be represented by a series voltage generator.

A long-lasting debate has been created by the "additional" noise of MOS transistors observed at high frequencies.[39] Repeated experiments and theoretical explanations have not always produce consistent results.[5, 40, 45, 70, 72, 111, 126, 131, 146, 176] In retrospect, it seems that the additional noise may have been generated in most cases by the hot electron effect;[133, 134] however, the condition for this recognition is the systematic investigation of the ionization in the field-effect transistor.[2, 68, 98, 114, 115, 131]

The fixed charges within the depletion layer, as well as the substrate doping, may also have an effect,[74, 117] although this effect is small in conventional structures.[110] In ion-implanted, sapphire grown (SOS) MOS field-effect transistors, a noise increase due to high-frequency mutual conductance reduction has been observed.[10, 57]

Concerning the low-frequency excess noise in MOS transistors, it is practically impossible to survey the published experimental results and frequently contradictory theoretical models. However, two observations seem to be generally true:

1. The spectrum practically always has a $1/f$ shape.
2. There is strong correlation between the noise power density and surface state density at the Si—SiO$_2$ interface.

Without trying to present a complete survey, some references are given comprising predominantly experimental results[1, 27, 29, 30, 35, 46, 48, 58, 92, 99, 172, 175] or theoretical results.[11, 18, 32, 49, 53, 62, 64, 73, 88, 148, 152, 154, 173, 174] Das and Moore[27] have provided a good survey on the principal theoretical trends. Other papers[24, 83, 113] have considered application problems and noise figure optimization procedures.

Noise sources of junction and MOS field-effect transistor devices are summarized in Fig. 6-29 with the aid of noise sources reduced to the input terminal. Subscripts t, f and G denote the components originating from the thermal effect, flicker effect, and the gate electrode DC current. Voltage generators represent the channel noise, while the induced gate electrode noise is represented by current generators. There is a correlation between generators of identical subscripts. For MOS transistors $i_G=0$, and for junction field-effect transistors it may be calculated from Eq. (6-134).

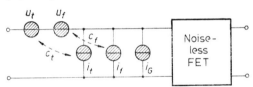

Figure 6-29 General noise equivalent
circuit of field-effect transistors.

PROBLEMS

6-1 State the device type and working point conditions in the low-frequency range, but covering the white-noise region, for signal sources resistances of $R_g=1$, 10, 100 kΩ. The available semiconductor devices are: Si n–p–n transistor—$\beta=300$ and $r_{bb'}=200$ Ω, and field-effect transistor—$g_{ms}=5$ mA/V and $I_G=10^{-9}$ A.

Answer: The optimum working point currents of the n–p–n transistor are, from Eq. (6-80), 375, 44, and 4.5 μA and the noise figures, from Eq. (6-82), are 1.275, 1.085, and 1.062. However, at a working point current of 4.5 μA, the current-gain factor probably decreases substantially [see Eq. (6-87)], so the extremely favorable value of 1.062 may not be achieved.

In case of the field-effect transistor, the source parameters of the low-frequency equivalent noise generators (see Fig. 6-27b) are almost independent of the working point in the pinch-off region. The channel noise is

$$u^2 = \frac{4kT\Delta f}{g_{ms}} K_d \qquad 0.5 < K_d < 0.666$$

and the gate electrode current noise is

$$i_g^2 = 2qg I_G \Delta f \qquad g \cong 1 \tag{6-134}$$

Applying the approximation $K_d \cong 0.6$ and utilizing Eq. (5-50), the following optimum source resistance pertains to the given values of u and i_g:

$$R_{g\,opt} = \frac{u}{i_g} = \sqrt{\frac{2U_T K_d}{g_m I_G}} = 79 \text{ k}\Omega$$

Here, according to Eq. (5-51) and assuming $a=0$, we get $F_{min}=1.003$. The noise figures pertaining to the signal source resistance of 1, 10, and 100 kΩ are 1.12, 1.012, and 1.003. This means that in all cases, the field-effect transistor is better than the bipolar transistor.

6-2 Assume that the current-gain factor of the bipolar transistor applied in Problem 6-1 is a function of the working point current according to Eq. (6-63) where $m=1.5$ and $B(375 \text{ }\mu\text{A})=200$. What is the optimum working point current and noise figure for $R_g=100$ kΩ?

Answer: It can easily be shown that (6-83) holds, so (6-86) may be approximated in a form similar to Eq. (6-84). Thus

$$F \cong 1 + \frac{r_e/2 + r_{bb'}}{R_g} + \frac{R_g m}{2\beta r_e}$$

In the last term,

$$\frac{\beta}{m} = B = kI_C^{1-1/m} \cong kI_E^{1-1/m}$$

where k can be calculated by substituting the given values into Eq. (6-63). Furthermore, $r_e = U_T/I_E$ and $r_{bb'} \ll r_e/2$ (the end result of this problem justifies this neglection), so

$$F = 1 + \frac{U_T}{2I_E R_g} + \frac{R_g}{2kU_T} I_E^{1/m}$$

This function has a minimum value for

$$I_E = \left(mk\frac{U_T^2}{R_g^2} \right)^{m/(m+1)} = 1.86 \times 10^{-6} \text{ A}$$

where

$$F_{min} = 1.14$$

6-3 The DC current flowing through the floating resistance of 100 Ω has to be measured with as low a noise figure as possible. Utilizing the bipolar transistor-type given in Problem 6-1, what is the suitable circuit?

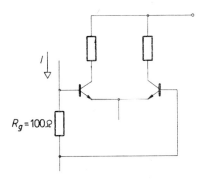

Figure P6-1 Differential amplifier
used for current measurement.

Answer: Both the floating resistance and the DC transmission require a differential amplifier (Fig. P6-1). According to the reasoning given in Sec. 6-5, half of the resistance of the signal source has to be taken into account for the noise calculation of half of the differential amplifier. In the given frequency range, flicker noise dominates, so according to (6-98) we have $R_{g\,opt}/2 = r_{bb'}$. As for the transistor type in question ($r_{bb'} = 200 \ \Omega$), four+four transistors have to be connected in parallel. It is advisable to adjust a working point current to as low a value as possible [see Eq. (6-98)].

REFERENCES

[1] Abowitz, G. *et al.*: "Surface States and $1/f$ Noise in MOS Transistors", *IEEE Trans.*, vol. ED-14, no. 11, pp. 775—777, 1967.

[2] Ambrózy, A.: "On the Gate Current of Junction Field Effect Transistors", *Period. Polytechn.*, vol. 14, no. 4, pp. 355—361, 1970.

[3] — *et al.*: "Direct Reading Transistor Noise Figure Measuring Instrument", *Electr. Eng.*, vol. 35, no. 9, pp. 611—613, 1963.

[4] Anastassiou, A. and M. J. O. Strutt: "Effect of Source Load Inductance on the Noise Figure of GaAs FET", *Proc. IEEE*, vol. 62, no. 3, pp. 406—408, 1974.

[5] Bächtold, W.: "Noise Behaviour of GaAs Field-Effect Transistors with Short Gate Lengths", *IEEE Trans.*, vol. ED-19, no. 5, pp. 674—680, 1972.

[6] Bächtold, W. and M. J. O. Strutt: "Analogue Model for the Signal and Noise Equivalent Circuit of UHF Transistors", *Electr. Lett.*, vol. 2, no. 9, pp. 335—336, 1966.

[7] — and —: "Simplified Equivalent Circuit for the Noise Figure Calculation of Microwave Transistors", *Electr. Lett.*, vol. 4, no. 10, pp. 209—210, 1968.

[8] — and —: "Optimum Source Admittance for Minimum Noise Figure of Microwave Transistors", *Electr. Lett.*, vol. 4, no. 17, pp. 346—348, 1968.

⁹ Baelde, A.: "Theory and Experiments on the Noise of Transistors", *Philips Res. Rept. Suppl.*, vol. 20, no. 4, 1965.

¹⁰ Baril, W. A. *et. al*: "High-Frequency Thermal Noise in MOSFETs", *Solid State Electr.*, vol. 21, no. 3, pp. 589—592, 1978.

¹¹ Berz, F.: "Theory of Low Frequency Noise in Si MOSTs", *Solid State Electr.*, vol. 13, no. 5, pp. 631—647, 1970.

¹² Bilotti, A. and E. Mariani: "Noise Characteristics of Current Mirror Sinks/Sources", *IEEE J.*, vol. SC-10, no. 6, pp. 516—524, 1975.

¹³ Brodersen, A. J. *et al.*: "Noise in Integrated Circuit Transistors", *IEEE J.*, vol. SC-5, no. 2, pp. 63—66, 1970.

¹⁴ Bruncke, W. C. *et al.*: "Transistor Noise at Low Temperatures", *IEEE Trans.*, vol. ED-11, no. 2, pp. 50—53, 1964.

¹⁵ — and A. van der Ziel: "Thermal Noise in Junction Gate Field-Effect Transistors", *IEEE Trans.*, vol. ED-13, no. 3, pp. 323—329, 1966.

¹⁶ Buckingham, M. J. and E. A. Faulkner: "The Theory of Inherent Noise in p–n Junction Diodes and Bipolar Transistors", *Radio Electr. Eng.*, vol. 44, no. 3, pp. 125—140, 1974.

¹⁷ Chenette, E. R. and A. van der Ziel: "Accurate Noise Measurements on Transistors", *IRE Trans.*, vol. ED-9, no. 3, pp. 123—128, 1962.

¹⁸ Christensson, S. *et al.*: "Low Frequency Noise in MOS Transistors", *Solid State Electr.*, vol. 11, no. 9, pp. 797—820, 1968.

¹⁹ Conti, M.: "Surface and Bulk Effects in Low Frequency Noise in n–p–n Planar Transistors", *Solid State Electr.*, vol. 13, no. 11, pp. 1461—1469, 1970.

²⁰ — and G. Corda: "Noise Sources Identification in Integrated Circuits through Correlation Analysis", *IEEE J.*, vol. SC-9, no. 3, pp. 124—133, 1974.

²¹ Cook, K. B. and A. J. Brodersen: "Physical Origins of Burst Noise in Transistors", *Solid State Electr.*, vol. 14, no. 12, pp. 1237—1250, 1971.

²² Cowley, A. M. and R. A. Zettler: "Shot Noise in Silicon Schottky–Barrier Diodes", *IEEE Trans.*, vol. ED-15, no. 10, pp. 761—769, 1968.

²³ Das, M. B.: "Generalized High-Frequency Network Theory of Field Effect Transistors", *Proc. IEE*, vol. 114, no. 1, pp. 50—59, 1967.

²⁴ —: "FET Noise Sources and Their Effects on Amplifier Performance at Low Frequencies", *IEEE Trans.*, vol. ED-19, no. 3, pp. 338—348, 1972.

²⁵ —: "On the Current Dependence of Low-Frequency Noise in Bipolar Transistors", *IEEE Trans.*, vol. ED-22, no. 12, pp. 1092—1098, 1975.

²⁶ — and O. A. Dogha: "On the Noise Performance of Bipolar Transistors in Untuned Amplifiers", *Solid State Electr.*, vol. 19, no. 10, pp. 827—836, 1976.

²⁷ — and J. M. Moore: "Measurements and Interpretation of Low-Frequency Noise in FETs", *IEEE Trans.*, vol. ED-21, no. 4, pp. 247—257, 1974.

²⁸ Faulkner, E. A.: "The Design of Low-Noise Audio Frequency Amplifiers", *Radio Electr. Eng.*, vol. 36, no. 1, pp. 17—30, 1968.

²⁹ Fichtner, W. and E. Hochmair: "Current-Kink Noise of n-Channel Enhancement ESFI MOS SOS Transistors", *Electr. Lett.*, vol. 13, no. 22, pp. 675—676, 1977.

³⁰ Flinn, I. *et al.*: "Low-Frequency Noise in MOS Field-Effect Transistors", *Solid State Electr.*, vol. 10, no. 8, pp. 833—846, 1967.

³¹ Fu, H. S. and C. T. Sah: "Lumped Model Analysis of the Low-Frequency Generation Noise in JFETs", *Solid State Electr.*, vol. 12, no. 8, pp. 605—618, 1969.

³² — and —: "Theory and Experiments on Surface $1/f$ Noise", *IEEE Trans.*, vol. ED-19, no. 2, pp. 273—285, 1972.

³³ Fukui, H.: "The Noise Performance of Microwave Transistors", *IEEE Trans.*, vol. ED-13, no. 3, pp. 329—341, 1966.

[34] Fuse, N. *et al.*: "Low-Noise Transistors", *Toshiba Review*, no. 56, pp. 22—26, April, 1971.

[35] Gentil, P. and S. Chausse: "Low-Frequency Measurement on Silicon-on-Sapphire (SOS) MOS Transistors", *Solid State Electr.*, vol. 20, no. 11, pp. 935—940, 1977.

[36] Gibbons, J. F.: "Low-Frequency Noise Figure and Its Application to the Measurement of Certain Transistor Parameters", *IRE Trans.*, vol. ED-9, no. 5, pp. 308—315, 1962.

[37] Grove, A. S. and D. J. Fitzgerald: "Surface Effects on p–n Junctions", *Solid State Electr.*, vol. 9, no. 8, pp. 783—806, 1966.

[38] Guggenbühl, W. and M. J. O. Strutt: "Theory and Experiments of Shot Noise in Semiconductor Junction Diodes and Transistors", *Proc. IRE*, vol. 45, no. 6, pp. 839—854, 1957.

[39] Halladay, H. E. and A. van der Ziel: "Test of the Thermal Noise Hypothesis in MOSFETs", *Electr. Lett.*, vol. 4, no. 17, pp. 366—367, 1968.

[40] — and —: "On the High-Frequency Excess Noise and Equivalent Circuit Representation of the MOSFET with n-Type Channel", *Solid State Electr.*, vol. 12, no. 3, pp. 161—176, 1969.

[41] Haslett, J. W.: "Noise Performance of the New Norton Op Amps", *IEEE Trans.*, vol. ED-21, no. 9, pp. 571—577, 1974.

[42] — and E. J. M. Kendall: "Temperature Dependence of Low-Frequency Excess Noise in Junction Gate FETs", *IEEE Trans.*, vol. ED-19, no. 8, pp. 943—950, 1972.

[43] — and F. N. Trofimenkoff: "Generation Noise Resistance in Junction Field-Effect Transistors at Pinch-Off", *Solid State Electr.*, vol. 12, no. 9, pp. 747—749, 1969.

[44] — and —: "Thermal Noise in Field-Effect Devices", *Proc. IEE*, vol. 116, no. 11, pp. 1863—1868, 1969.

[45] — and —: "Gate Noise in MOSFET's at Moderately High Frequencies", *Solid State Electr.*, vol. 14, no. 3, pp. 239—245, 1971.

[46] — and —: "Effects of the Substrate on Surface State Noise in Silicon MOSFETs", *Solid State Electr.*, vol. 15, no. 1, pp. 117—131, 1972.

[47] Hawkins, R. J.: "Limitations of Nielsen's and Related Noise Equations", *Solid State Electr.*, vol. 20, no. 3, pp. 191—196, 1977.

[48] — and G. G. Bloodworth: "Two Components of $1/f$ Noise in MOS Transistors", *Solid State Electr.*, vol. 14, no. 10, pp. 932—939, 1971.

[49] Hayashi, A. and A. van der Ziel: "Correlation Coefficient of Gate and Drain Flicker Noise", *Solid State Electr.*, vol. 17, no. 6, pp. 637—639, 1974.

[50] Hidas, G. and A. Ambrózy: "Examination of Surface Defects of Semiconductor Devices by Noise Measurements", *2nd Symposium on Reliability in Electronics*, Budapest, 1968, paper no. 212.

[51] Hsu, S. T.: "Surface State-Related $1/f$ Noise in p–n Junctions", *Solid State Electr.*, vol. 13, no. 6, pp. 843—855, 1970.

[52] —: "Low-Frequency Excess Noise in Metal-Silicon Schottky–Barrier Diodes", *IEEE Trans.*, vol. ED-17, no. 7, pp. 496—506, 1970.

[53] —: "Surface State-Related $1/f$ Noise in MOS Transistors", *Solid State Electr.*, vol. 13, no. 11, pp. 1451—1459, 1970.

[54] —: "Noise in High-Gain Transistors and Its Application to the Measurement of Certain Transistor Parameters", *IEEE Trans.*, vol. ED-18, no. 7, pp. 425—431, 1971.

[55] —: "Flicker Noise in Metal Semiconductor Schottky–Barrier Diodes due to Multistep Tunneling Process", *IEEE Trans.*, vol. ED-18, no. 10, pp. 882—887, 1971.

[56] —: "Bistable Noise in Operational Amplifiers", *IEEE J.*, vol. SC-6, no. 6, pp. 399—403, 1971.

[57] — and A. van der Ziel: "Thermal Noise in Ion-Implanted MOSFETs", *Solid State Electr.*, vol. 18, no. 6, pp. 509—510, 1975.

58 — and —: "A New Type of Flicker Noise in Microwave MOSFETs", *Solid State Electr.*, vol. 18, no. 10, pp. 885—886, 1975.

59 Huang, R. S. and P. H. Ladbrooke: "The Physics of Excess Electron Velocity in Submicron Channel FET's", *J. Appl. Phys.*, vol. 48, no. 11, pp. 4791—4798, 1977.

60 Jaeger, R. C. and A. J. Brodersen: "Low-Frequency Noise Sources in Bipolar Junction Transistors", *IEEE Trans.*, vol. ED-17, no. 2, pp. 128—134, 1970.

61 Jäntsch, O. and I. Feigt: 1/f Noise in Silicon Diodes", *Solid State Electr.*, vol. 16, no. 12, pp. 1517—1520, 1973.

62 Jindal, R. P. and A. van der Ziel: "Carrier Fluctuation Noise in a MOSFET Channel due to Traps in the Oxide", *Solid State Electr.*, vol. 21, no. 6, pp. 901—903, 1978.

63 Johnson, K. H. *et al.*: "Transistor Noise at High Injection Levels", *IEEE Trans.*, vol. ED-12, no. 6, pp. 387—388, 1965.

64 Jordan, A. G. and N. A. Jordan: "Theory of Noise in Metal Oxide Semiconductor Devices", *IEEE Trans.*, vol. ED-12, no. 3, pp. 148—156, 1965.

65 Kandiah, K. and F. B. Whiting: "Low-Frequency Noise in Junction Field-Effect Transistors", *Solid State Electr.*, vol. 21, no. 8, pp. 1079—1088, 1978.

66 Kässer, R.: "A New Noise Equivalent Circuit for the Junction FET", *Proc. IEEE*, vol. 58, no. 7, pp. 1171—1172, 1970.

67 —: "Noise Factor Contours for Field-Effect Transistors at Moderately High Frequencies", *IEEE Trans.*, vol. ED-19, no. 2, pp. 164—171, 1972.

68 Kim, C. S.: "Avalanche Multiplication and Related Noise in Silicon MOSFETs", Ph. D. Thesis, University of Florida, 1971.

69 Kingston, R. H. (ed.): *Semiconductor Surface Physics*, University of Pennsylvania Press, Philadelphia, 1957.

70 Kirk, E. W. *et al.*: "Induced Gate Noise in MOSFETs", *Solid State Electr.*, vol. 14, no. 10, pp. 945—948, 1971.

71 Klaassen, F. M.: "High-Frequency Noise of the Junction Field-Effect Transistor", *IEEE Trans.*, vol. ED-14, no. 7, pp. 368—373, 1967.

72 —: "Comments on Hot-Carrier Noise in Field-Effect Transistors", *IEEE Trans.*, vol. ED-18, no. 1, pp. 74—75, 1971.

73 —: "Characterization of Low 1/f Noise in MOS Transistors", *IEEE Trans.*, vol. ED-18, no. 10, pp. 887—891, 1971.

74 — and J. Prins: "Thermal Noise of MOS Transistors", *Philips Res. Rept.*, vol. 22, no. 5, pp. 505—514, 1967.

75 — and —: "Noise of Field-Effect Transistors at Very High Frequencies", *IEEE Trans.*, vol. ED-16, no. 11, pp. 952—957, 1969.

76 Kleinpenning, T. G. M.: "Low Frequency Noise in Schottky–Barrier Diode", *Solid State Electr.*, vol. 22, no. 2, pp. 121—128, 1979.

77 Knott, K. F.: "Comparison of Varactor Diode and Junction FET Low Noise LF Amplifiers", *Electr. Lett.*, vol. 3, no. 11, p. 512, 1967.

78 —: "Experimental Location of the Surface and Bulk 1/f Noise Currents in Transistors", *Solid State Electr.*, vol. 16, no. 12, pp. 1429—1434, 1973.

79 —: "Comments on the Emitter Current Density Dependence of Popcorn Noise Frequency", *IEEE Trans.*, vol. ED-25, no. 4, pp. 494—495, 1978.

80 Koji, T.: "The Effect of Emitter Current Density on Popcorn Noise in Transistors", *IEEE Trans.*, vol. ED-22, no. 1, pp. 24—25, 1975.

81 Lauritzen, P. O.: "Low-Frequency Generation Noise in Junction Field-Effect Transistors", *Solid State Electr.*, vol. 8, no. 1, pp. 41—58, 1965.

82 —: "Noise due to Generation and Recombination of Carriers in p–n Junction Transition Regions", *IEEE Trans.*, vol. ED-15, no. 10, pp. 770—776, 1968.

[83] Lecoy, G. *et al.*: "Equivalent Noise Generators in JG FET", *Solid State Electr.*, vol. 17, no. 1, pp. 11—16, 1974.

[84] Lee, S. J. and A. van der Ziel: "Flicker-Noise Compensation in High Impedance MOSFET Circuits", *Solid State Electr.*, vol. 16, no. 11, pp. 1301—1302, 1973.

[85] Leuenberger, F.: "1/f Noise in Gate-Controlled Planar Silicon Diodes", *Electr. Lett.*, vol. 4, no. 13, p. 280, 1968.

[86] Leupp, A. and M. J. O. Strutt: "Noise Behaviour of the MOSFET at VHF and UHF", *Electr. Lett.*, vol. 4, no. 15, pp. 313—314, 1968.

[87] — and —: "High-Frequency FET Noise Parameters and Approximation of the Optimum Source Admittance", *IEEE Trans.*, vol. ED-16, no. 5, pp. 428—431, 1969.

[88] Leventhal, E. A.: "Derivation of 1/f Noise in Silicon Inversion Layers from Carrier Motion in a Surface Band", *Solid State Electr.*, vol. 11, no. 6, pp. 621—627, 1968.

[89] Lindmayer, J. and C. Y. Wrigley: *Fundamentals of Semiconductor Devices*, D. Van Nostrand, Princeton, 1965.

[90] López de la Fuente, J.: "Measurement of Very Low-Frequency Noise", Ph. D. Thesis, University of Eindhoven, 1970.

[91] Malaviya, S. D. and A. van der Ziel: "A Simplified Approach to Noise in Microwave Transistors", *Solid State Electr.*, vol. 13, no. 12, pp. 1511—1518, 1970.

[92] Mansour, I. R. M. *et al.*: "Physical Model for the Current Noise Spectrum of MOSTs", *Brit. J. Appl. Phys.*, vol. 2, no. 7, pp. 1063—1082, 1969.

[93] Martin, J. C. *et al.*: "Mesures du bruit de fond des transistors plans aux très basses frequences", *Electr. Lett.*, vol. 2, no. 9, pp. 343—345, 1966.

[94] McDonald, B. A.: "Avalanche Induced 1/f Noise in Bipolar Transistors", *IEEE* Trans., vol. ED-17, no. 2, pp. 134—136, 1970.

[95] Min, H. S. and K. M. van Vliet: "Theory of Intermediate and High-Injection Noise in Transistors", *Solid State Electr.*, vol. 17, no. 3, pp. 285—300, 1974.

[96] Müller, O.: "Temperature Fluctuations and Flicker Noise in p–n Junction Diodes", *IEEE Trans.*, vol. ED-21, no. 8, pp. 539—540, 1974.

[97] —: "A Formula for 1/f Flicker Noise in p–n Junctions; Calculating Flicker Noise of p–n Junction Diodes", *AEÜ*, vol. 28, no. 10—11, pp. 429—432, pp. 450—454, 1974.

[98] Nakahara, M. *et al.*: "Anomalous Low-Frequency Noise Enhancement in Silicon MOS Transistors; On the Gate Current and Noise Behaviour in JFETs", *Proc. IEEE*, vol. 57, no. 12, pp. 2177—2178, 1969; and vol. 58, no. 7, pp. 1158—1159, 1970.

[99] Nakamura, K. *et al.*: "Noise Characteristics of Ion-Implanted MOS Transistors", *J. Appl. Phys.*, vol. 46, no. 7, pp. 3189—3193, 1975.

[100] Neudeck, G. W.: "High-Frequency Shot Noise in Schottky–Barrier Diodes", *Solid State Electr.*, vol. 13, no. 9, pp. 1249—1256, 1970.

[101] — *et al.*: "The Ideality Factor and the High-Frequency Noise of Schottky–Barrier-Type Diodes", *IEEE J.*, vol. SC-7, no. 1, pp. 89—90, 1972.

[102] Nielsen, E. G.: "Behaviour of Noise Figure in Junction Transistors", *Proc. IRE*, vol. 45, no. 7, pp. 957—963, 1957.

[103] Nussbaum, A.: "Generation–Recombination Characteristic Behaviour of Silicon Diodes", *Phys. St. Sol.*, vol. A19, no. 2, pp. 441—450, 1973.

[104] Patterson, J. D.: "A Noise Model for the Distributed Transistor", *IEEE J.*, vol. SC-4, no. 2, pp. 75—80, 1969.

[105] Paul, R.: "Thermisches Rauschen von MOS Transistoren", *Nachrichtentechnik*, vol. 17, no. 12, pp. 458—466, 1967.

[106] Plumb, J. L. and E. R. Chenette: "Flicker Noise in Transistors", *IEEE Trans.*, vol. ED-10, no. 5, pp. 304—308, 1963.

[107] Pucel, R. A. *et al.*: "Signal and Noise Properties of Gallium Arsenide Microwave Field-

Effect Transistors", in L. Marton (ed.) *Advances in Electronic and Electron Physics*, vol. 38, Academic Press, New York, 1975.

[108] — *et al.*: "Noise Performance of Gallium Arsenide Field-Effect Transistors", *IEEE J.*, vol. SC-11, no. 2, pp. 243—255, 1976.

[109] Ram, G. V. and M. S. Tyagi: "A Quasi-One-Dimensional Analysis of Small Signal Current Gains in Lateral Transistors", *IEEE Trans.*, vol. ED-25, no. 11, pp. 1283—1290, 1978.

[110] Rao, P. S.: "The Effect of the Substrate upon the Gate and Drain Noise Parameters of MOSFETs", *Solid State Electr.*, vol. 12, no. 7, pp. 549—556, 1969.

[111] — and A. van der Ziel: "Noise and y Parameters in MOSFETs", *Solid State Electr.*, vol. 14, no. 10, pp. 939—944, 1971.

[112] Robinson, F. N. H.: "Noise in Junction Diodes and Bipolar Transistors at Moderately High Frequencies", *Electr. Eng.*, vol. 41, no. 2, pp. 218—220, 1969.

[113] —: "Noise in Common Source FET Amplifiers at Moderately High Frequencies", *Electr. Eng.*, vol. 41, no. 5, pp. 77—79, 1969.

[114] Rucker, L. M. and A. van der Ziel: "Noise Associated with JFET Gate Current Resulting from Avalanching in the Channel", *Solid State Electr.*, vol. 21, no. 5, pp. 798—799, 1978.

[115] — *et al.*: "On the Gate to Drain Ratio in Junction FETs at Low Temperature", *Solid State Electr.*, vol. 21, no. 3, pp. 596—597, 1978.

[116] Sah, C. T.: "Theory of Low-Frequency Generation Noise in Junction Gate Field-Effect Transistors", *Proc. IEEE*, vol. 52, no. 7, pp. 795—814, 1964.

[117] — *et al.*: "Carrier Generation and Recombination in p–n Junctions and p–n Junction Characteristics", *Proc. IRE*, vol. 45, no. 9, pp. 1228—1243, 1957.

[118] — *et al.*: "The Effect of Fixed Bulk Charge on the Thermal Noise in MOS Transistors", *IEEE Trans.*, vol. ED-13, no. 4, pp. 410—414, 1966.

[119] Schneider, B. and M. J. O. Strutt: "Theory and Experiments on Shot Noise in Si p–n Junctions and Transistors", *Proc. IRE*, vol. 47, no. 4, pp. 546—554, 1959.

[120] — and —: "Shot and Thermal Noise in Transistors at High-Level Current Injections", *Proc. IRE*, vol. 48, no. 10, pp. 1731—1739, 1960.

[121] Scott, L. and M. J. O. Strutt: "Spontaneous Fluctuations in the Leakage Current", *Solid State Electr.*, vol. 9, no. 10, pp. 1067—1073, 1966.

[122] Seltz, D. and I. Kidron: "A Two-Dimensional Model for the Lateral p–n–p Transistor", *IEEE Trans.*, vol. ED-21, no. 9, pp. 587—592, 1974.

[123] Shoji, M.: "Analysis of High-Frequency Thermal Noise of Enhancement Mode MOSFETs", *IEEE Trans.*, vol. ED-13, no. 6, pp. 520—524, 1966.

[124] Smulders, W.: "Noise Properties of Transistors at High Frequency", *Electr. Appl.*, vol. 23, no. 1, pp. 1—25, 1962.

[125] Smullin, L. D. and H. A. Haus: *Noise in Electron Devices*, John Wiley, New York, 1959.

[126] Sodini, D. *et al.*: "Impedance and Noise in the Channel of a GaAs Schottky-Gate Field-Effect Transistor", *Solid State Electr.*, vol. 20, no. 7, pp. 579—581, 1977.

[127] Solomon, J. E.: "Cascade Noise Figure of Integrated Transistor Amplifiers", *Motorola Application Note*, AN 223.

[128] Spenke, E.: "Das induktive Verhalten von p–n Gleichrichtern bei starken Durchlassbelastungen", *Z. Angew. Phys.*, vol. 10, no. 1, pp. 65—68, 1958.

[129] Statz, H. *et al.*: "Noise Characteristics of Gallium Arsenide Field-Effect Transistors", *IEEE Trans.*, vol. ED-21, no. 9, pp. 549—562, 1974.

[130] Stojadinovic, N. D.: "Effects of Emitter Edge Dislocations on the Low-Frequency Noise of Silicon Planar n–p–n Transistors", *Electr. Lett.*, vol. 15, no. 12, pp. 340—342, 1979.

[131] Takagi, K. and K. Matsumoto: "Noise in Silicon and FET's at High Electric Fields", *Solid State Electr.*, vol. 20, no. 1, pp. 1—3, 1977.

[132] — and A. van der Ziel: "Non-thermal Noise in MOSFETs and MOS Tetrodes", *Solid State Electr.,* vol. 12, no. 11, p. 907, 1969.

[133] Takagi, A. and A. van der Ziel: "High-Frequency Excess Noise and Flicker Noise in GaAs FETs", *Solid State Electr.,* vol. 22, no. 3, pp. 285—287, 1979.

[134] Takagi, K. and A. van der Ziel: "Excess High Frequency Noise and Flicker Noise in MOSFETs", *Solid State Electr.,* vol. 22, no. 3, pp. 289—292, 1979.

[135] Thommen, W. and M. J. O. Strutt: "Noise Figure of UHF Transistors", *IEEE Trans.,* vol. ED-12, no. 9, pp. 499—500, 1965.

[136] Tong, A. H. and A. van der Ziel: "Transistor Noise at High Injection Levels", *IEEE Trans.,* vol. ED-15, no. 5, pp. 307—313, 1968.

[137] Uhlir, A.: "High-Frequency Shot Noise in p–n Junctions", *Proc. IRE,* vol. 44, no. 4, pp. 557—558, 1956.

[138] Valkó, I. P.: "Untersuchung des Rauschens von Halbleitern", *Period. Polytechn.,* vol. 5, no. 1, pp. 57—73, 1961.

[139] van der Ziel, A.: *Noise,* Prentice Hall, Englewood Cliffs, 1954.

[140] —: "Noise in Junction Transistors", *Proc. IRE,* vol. 46, no. 6, pp. 1019—1038, 1958.

[141] —: *Fluctuation Phenomena in Semiconductors,* Butterworth, London, 1959.

[142] —: "Thermal Noise in Field-Effect Transistors", *Proc. IEEE,* vol. 50, no. 8, pp. 1808—1812, 1962.

[143] —: "Gate Noise in Field Effect Transistors at Moderately High Frequencies", *Proc. IEEE,* vol. 51, no. 3, pp. 461—467, 1963.

[144] —: "Equivalence of the Noise Figures of Common Source and Common Gate FET Circuits", *Electr. Lett.,* vol. 5, no. 8, pp. 161—162, 1969.

[145] —: *Noise: Sources, Characterization, Measurement,* Prentice Hall, Englewood Cliffs, 1970.

[146] —: "Noise Resistance of FETs in the Hot Electron Regime", *Solid State Electr.,* vol. 14, no. 4, pp. 347—350, 1971.

[147] —: "Shot Noise in Back Biased p–n Silicon Diodes", *Solid State Electr.,* vol. 18, no. 11, pp. 969—970, 1975.

[148] —: "Limiting Flicker Noise in MOSFETs", *Solid State Electr.,* vol. 18, no. 11, p. 1031, 1975.

[149] —: "Enhanced Shot Noise in p–n Diodes due to Thermal Feedback", *IEEE Trans.,* vol. ED-22, no. 10, pp. 964—965, 1975.

[150] —: *Solid State Physical Electronics,* Prentice Hall, Englewood Cliffs, 1976.

[151] —: *Noise in Measurements,* Wiley Interscience, New York, 1976.

[152] —: "Dependence of Flicker Noise in MOSFETs on Geometry", *Solid State Electr.,* vol. 20, no. 3, p. 267, 1977.

[153] —: "High Injection Noise in Transistors", *Solid State Electr.,* vol. 20, no. 8, pp. 715—720, 1977.

[154] —: "Some General Relationships for Flicker Noise in MOSFETs", *Solid State Electr.,* vol. 21, no. 4, pp. 623—624, 1978.

[155] — and A. G. T. Becking: "Theory of Junction Diode and Junction Transistor Noise", *Proc. IRE,* vol. 46, no. 3, pp. 589—594, 1958.

[156] — and J. W. Ero: "Small-Signal, High-Frequency Theory of Field-Effect Transistors", *IEEE Trans.,* vol. ED-11, no. 4, pp. 128—135, 1964.

[157] — and K. M. van Vliet: "Transmission Line Model of High Injection Noise in Junction Diodes", *Solid State Electr.,* vol. 20, no. 8, pp. 721—723, 1977.

[158] — et al.: "A More Accurate Expression for the Noise Figure of Transistors", *Solid State Electr.,* vol. 19, no. 2, pp. 149—151, 1976.

[159] van Vliet, K. M.: "Noise and Admittance of the Generation–Recombination Current of Junction Devices", *IEEE Trans.,* vol. ED-23, no. 11, pp. 1236—1246, 1976.

[160] van Vliet, K. M.: "The Transfer Impedance Method for Noise in Field-Effect Transistors", *Solid State Electr.*, vol. 22, no. 3, pp. 233—236, 1979.

[161] — and A. van der Ziel: "Physical Interpretation of Noise Reduction for the Generation-Recombination Current", *IEEE Trans.*, vol. ED-24, no. 8, pp. 1127—1129, 1977.

[162] Verhagen, C. M.: "Minimum Noise Setting of Transistors", *Proc. IEEE*, vol. 54, no. 1, pp. 83—84, 1966.

[163] Verster, T. C.: "Anomalies in Transistor Low-Frequency Noise", *Proc. IEEE*, vol. 55, no. 7, pp. 1204—1205, 1967.

[164] Viola, T. J. and R. J. Mattauch: "High-Frequency Noise in Schottky–Barrier Diodes", *Proc. IEEE*, vol. 61, no. 3, p. 393, 1973.

[165] Wade, T. E. and A. van der Ziel: "Noise in Bipolar Junction Transistors at High Injection Levels", *Solid State Electr.*, vol. 19, no. 5, pp. 381—383, 1976.

[166] — and —: "Noise Associated with Recombination in the Emitter Space Charge Region of Transistors", *Solid State Electr.*, vol. 19, no. 11, pp. 909—910, 1976.

[167] — et al.: "Noise Effects in Bipolar Junction Transistors at Cryogenic Temperatures: Part I", *IEEE Trans.*, vol. ED-23, no. 9, pp. 998—1007, 1976.

[168] — et al.: "Noise Effect in Bipolar Junction Transistors at Cryogenic Temperatures: Part II", *IEEE Trans.*, vol. ED-23, no. 9, pp. 1007—1011, 1976.

[169] Wall, E. L.: "Edge Injection Currents and their Effects on $1/f$ Noise in Planar Schottky Diodes", *Solid State Electr.*, vol. 19, no. 5, pp. 389—396, 1976.

[170] Wallmark, J. T. and H. Johnson: *Field-Effect Transistors*, Prentice Hall, Englewood Cliffs, 1966.

[171] Walton, C. A. and C. C. Liu: "A Low Noise Amplifier with Parallel Integrated Circuit Transistors", *IEEE J.*, vol. SC-6, no. 6, pp. 415—417, 1971.

[172] Wang, K. L.: "Measurements of Residual Defects and $1/f$ Noise in Ion-Implanted p-Channel MOSFETs", *IEEE Trans.*, vol. ED-25, no. 4, pp. 478—484, 1978.

[173] Wu, S. Y.: "Theory of the Generation–Recombination Noise in MOS Transistors", *Solid State Electr.*, vol. 11, no. 1, pp. 25—32, 1968.

[174] Yau, L. D. and C. T. Sah: "Low-Frequency Generation–Recombination Noise in MOS Transistors", *IEEE Trans.*, vol. ED-16, no. 2, pp. 170—177, 1969.

[175] — and —: "Geometrical Dependences of the Low-Frequency Generation–Recombination Noise in MOS Transistors", *Solid State Electr.*, vol. 12, no. 11, pp. 903—905, 1969.

[176] — and —: "On the Excess 'White Noise' in MOS Transistors", *Solid State Electr.*, vol. 12, no. 12, pp. 927—936, 1969.

[177] "Hot Electrons in Semiconductors", *Solid State Electr.*, vol. 21, no. 1, pp. 5—320, 1978.

[178] Wolf, E. D. (ed.): *Noise in Physical Systems*, Springer, Heidelberg, 1978.

SEVEN

EFFECTS OF NONLINEAR TRANSFER
CHARACTERISTICS

In this Chapter, noise generation processes will not be investigated as in linear networks; the noise voltage or current is seldom high enough to preclude the application of the small-signal linear equivalent circuit. In spite of this, these networks have to be investigated because of their application to noise-measuring apparatus. According to previous Chapters, present theories are not suitable for describing all noise phenomena, so measurements are of great importance. The detector characteristics in measuring apparatus are not necessarily linear; further nonlinearity may be introduced by the finite driving level capability of the amplifier preceding the detector. This means that the change in a stochastic signal amplitude and frequency spectrum while passing through the nonlinear network has to be determined.[8, 20] Our results will be valid for nonlinear networks in general. For simplicity, input signals having zero mean value will be assumed.

7-1 TRANSFORMATION OF THE AMPLITUDE DENSITY FUNCTION

Let us consider the nonlinear two-port shown in Fig. 7-1. Let the transfer characteristic be

$$y = g(x) \tag{7-1}$$

which is a single-valued, monotonically increasing, or decreasing, real function. Equation (7-1) may be applied to describe any detector, x and/or y may denote either voltage or current. What will be the function $f_2(y)$ describing the amplitude density of the output signal $\eta(t)$ in the case of an input stochastic signal $\xi(t)$ having a density function $f_1(x)$? In this case, a further restriction is necessary: $\eta(t)$ should be able to follow $\xi(t)$ instantaneously, so the network may not contain delaying elements.

Let the possible values of the random variable ξ pertain to the event space x and the values of η pertain to y. If the above conditions are met, the transfer characteristic function $y=g(x)$ will provide a one-to-one mapping of x to y or vice versa. The probability of $\eta < y$ is

$$F_2(y) = \mathscr{P}(\eta < y) = \mathscr{P}[\xi < x(y)] = F_1(x) \tag{7-2}$$

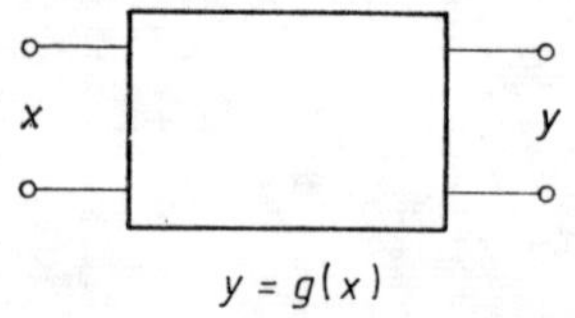

$$y = g(x)$$

Figure 7-1 Two-port with non-linear transfer function. x and y may be either voltage or current.

where $x(y)$ is the inverse of the mapping function (7-1). The relation between the density functions is derived utilizing Eq. (2-13). Thus

$$f_2(y) = \frac{d}{dy} F_2(y) = \frac{d}{dy} F_1[x(y)] = \frac{d}{dy} \int_{-\infty}^{x(y)} f_1(v)\, dv \tag{7-3}$$

where v is a temporary variable needed for performing the integration. After performing the indicated operations,

$$f_2(y) = f_1(x) \frac{dx}{dy} = \frac{f_1(x)}{g'(x)} \tag{7-4}$$

where $g'(x) = dy/dx$. Neither of the density functions may be negative so Eq. (7-4) will only yield a correct result if $g'(x) > 0$. However, it may be easily realized that the relationship

$$f_2(y) = f_1(x) \left| \frac{dx}{dy} \right| \tag{7-5}$$

is valid both for monotonically increasing and decreasing mapping functions.

Rectifier circuits usually exhibit a symmetrical characteristics as shown in Fig. 7-2. In this case for $y \geq 0$,

$$f_2(y) = [f_1(-x) + f_1(x)] \left| \frac{dx}{dy} \right| = 2f_1(x) \left| \frac{dx}{dy} \right| \tag{7-6}$$

According to the restriction of the introduction, the expected value of the input signal is zero. Also, let the density function be gaussian, and let us investigate the three characteristic transformations that are most important practically.

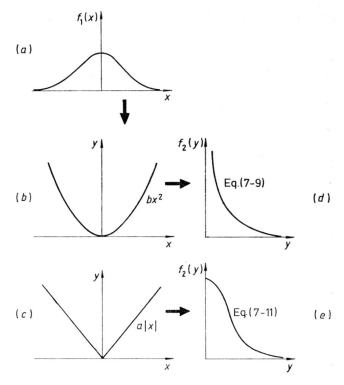

Figure 7-2 *(a)* Density function of the input signal. *(b)* Square-law detector characteristic. *(c)* Full-wave linear detector characteristic. *(d)* Density function of output signal for case *b* *(e)* Density function of output signal for case *c*.

First, let the detector characteristic be quadratic:

$$y = bx^2 \tag{7-7}$$

In this case,

$$\frac{dx}{dy} = \frac{1}{2\sqrt{by}} \tag{7-8}$$

and according to Eq. (7-6) $y \geq 0$, thus

$$f_2(y) = \frac{1}{\sqrt{by}} \frac{1}{\sigma\sqrt{2\pi}} e^{-x^2/2\sigma^2} = \frac{1}{\sigma\sqrt{2\pi by}} e^{-y/2b\sigma^2} \tag{7-9}$$

This is the χ^2 distribution of one degree of freedom given by Eq. (2-116). Thus the distribution at the detector output differs significantly from the normal distribution.

The equation of the characteristic pertaining to a full-wave, piecewise linear detector is

$$y = a|x| \tag{7-10}$$

The density function of the output signal then has the form

$$f_2(y) = \frac{2}{a\sigma\sqrt{2\pi}} e^{-y^2/2a^2\sigma^2} \tag{7-11}$$

and its domain is again $y \geq 0$.

The output signal density function for a half-wave detector having piecewise linear characteristics is somewhat more complicated (see Fig. 7-3). The equation of the characteristic is

$$y = 1(x)ax \tag{7-12}$$

where $1(x)$ is the unit step function. It is evident that

$$\mathscr{P}(\eta < y) = 0 \qquad \text{if} \qquad y < 0$$

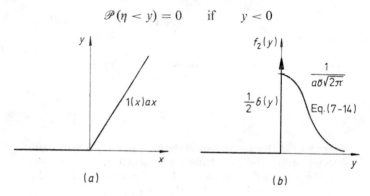

Figure 7-3 (a) Half-wave linear detector characteristic. (b) Density function of output signal for input signal of normal distribution.

On the other hand, the distribution function pertaining to positive y values should comprise the mapping of the points pertaining to the negative region $x<0$ of the distribution function $f_1(x)$. According to Eq. (7-2),

$$F_2(y) = \mathscr{P}(\eta < y) = \mathscr{P}(x < 0) + \int_0^{x(y)} f_1(v)\, dv \tag{7-13}$$

Performing the derivative according to Eq. (7-3), for a normal input distribution

$$f_2(y) = \frac{1}{2}\delta(y) + \frac{1(y)}{a\sigma\sqrt{2\pi}} e^{-y^2/2a^2\sigma^2} \tag{7-14}$$

where $\delta(y)$ is the Dirac-delta function defined in Eq. (3-78). This result may easily be visualized. The probability of an output signal with zero instantaneous value is very high as the output is zero for negative instantaneous values of the input signal that have 50 percent probability.

The first- and second-order moments of the density functions (7-9), (7-11), and (7-14) to be calculated below will yield the expectation and the variance of the detected signal.

7-2 TRANSFORMATION OF THE POWER SPECTRUM

It has been shown in Sec. 3-5 that the power spectrum of a signal passing through a linear network changes, as does the equivalent Fourier transform—the auto-correlation function. Nonlinear networks have a similar but more complicated effect, so our investigations will be restricted to special cases.

Let the equation of the nonlinear network characteristic be given by the power function

$$y = g(x) = cx^n \tag{7-15}$$

where n is a positive integer. A stationary stochastic input signal $\xi(t)$ with zero expectation generates, without delay, an output signal $\eta(t)$ that has the auto-correlation function

$$\Gamma_\eta(\tau) = \frac{1}{2T} \int_{-T}^{T} c^2 \xi^n(t) \xi^n(t+\tau)\, dt = c^2 M(\xi_t^n \xi_{t+\tau}^n) \tag{7-16}$$

Let the normalized autocorrelation function of the input signal be

$$\gamma_\xi(\tau) = \frac{\Gamma_\xi(\tau)}{M[\xi^2(t)]} = \frac{M[\xi(t)\xi(t+\tau)]}{M[\xi^2(t)]} \tag{7-17}$$

$\gamma_\xi(t)$ thus introduced is suitable for expressing the relationship between the instantaneous values of $\xi(t+\tau)$ and $\xi(t)$:[30]

$$\xi(t+\tau) = \gamma_\xi(\tau)\xi(t) + v(t) \tag{7-18}$$

where $v(t)$ is a random variable independent of $\xi(t)$ and with zero expectation, referring to the nondeterministic nature of the relation. The expectation of the squared value of Eq. (7-18) is

$$M[\xi^2(t+\tau)] = \gamma_\xi^2(\tau) M[\xi^2(t)] + M[v^2(t)] \tag{7-19}$$

because according to Eq. (2-21) and because $M[\xi(t)] = M[v(t)] = 0$

$$M[\xi(t)v(t)] = M[\xi(t)]M[v(t)] = 0 \tag{7-20}$$

However, the left-hand side may be substituted by $M[\xi^2(t)]$ as the process is stationary. Thus

$$M[v^2(t)] = [1 - \gamma_\xi^2(\tau)] M[\xi^2(t)] \tag{7-21}$$

So everything needed is available to calculate Eq. (7-16). Substituting Eq. (7-18) yields

$$\Gamma_\eta(\tau) = c^2 M\{\xi^n(t)[\gamma_\xi(\tau)\xi(t)+v(t)]^n\} \tag{7-22}$$

which may be evaluated utilizing expressions (7-19), (7-20), and (7-21).

In the case of most importance in practice, $n=2$ and the equation of the characteristic is utilized in the form of Eq. (7-7). Then

$$M[\xi^2(t)\xi^2(t+\tau)] = M(\xi^4)\gamma_\xi^2(\tau)+M(\xi^2)M(\xi^2)[1-\gamma_\xi^2(\tau)] \tag{7-23}$$

If $\xi(t)$ has normal distribution and a standard deviation of σ, the simple relation

$$\Gamma_\eta(\tau) = b^2[3\sigma^4\gamma_\xi^2(\tau)+\sigma^4-\sigma^4\gamma_\xi^2(\tau)] = b^2\sigma^4+2b^2\Gamma_\xi^2(\tau) \tag{7-24}$$

is derived utilizing Eqs. (2-72), (2-74), and (7-17). Equation (7-22) is also valid for positive integers greater than two, although results are much more complicated. For power relationships with fractional exponents and for other types of relationships, transformation properties are described using procedures based on Fourier or Laplace transforms.[3,8]

The power spectrum is the Fourier transform of $\Gamma_\eta(\tau)$. In most practical cases, $\Gamma_\eta(\tau)$ is at most a second-degree function of $\Gamma_\xi(\tau)$ [e.g., see Eq. (7-24)]. We therefore need the Fourier transform of $\Gamma_\xi^2(\tau)$:

$$F_{\Gamma^2}(\omega) = \int_{-\infty}^{\infty} \Gamma_\xi^2(\tau)e^{-j\omega\tau}\,d\tau = \int_{-\infty}^{\infty} \Gamma_\xi(\tau)[\Gamma_\xi(\tau)e^{-j\omega\tau}]\,d\tau \tag{7-25}$$

According to Eq. (3-63)

$$\Gamma_\xi(\tau) = \int_{-\infty}^{\infty} s(\Omega)e^{j\Omega\tau}\,d\Omega \tag{7-26}$$

Substituting this into Eq. (7-25), we have

$$F_{\Gamma^2}(\omega) = \iint_{-\infty}^{\infty} s(\Omega)e^{j(\Omega-\omega)\tau}\Gamma_\xi(\tau)\,d\Omega\,d\tau$$

$$= \iint_{-\infty}^{\infty} s(\Omega)\Gamma_\xi(\tau)e^{-j(\omega-\Omega)\tau}\,d\tau\,d\Omega = 2\pi\int_{-\infty}^{\infty} s(\Omega)s(\omega-\Omega)\,d\Omega \tag{7-27}$$

where Eq. (3-64) has been utilized.

Equation (7-27) shows that in the case of an input signal with a continuous power spectrum, the detector output signal will be the resultant of an infinite number of mixing products of different frequencies ω. Figure 7-4 shows the derivation of the power spectrum pertaining to an arbitrarily fixed value of ω. For an input signal having a discrete spectrum, Eq. (7-27) will give the results well known from the deterministic signal theory.

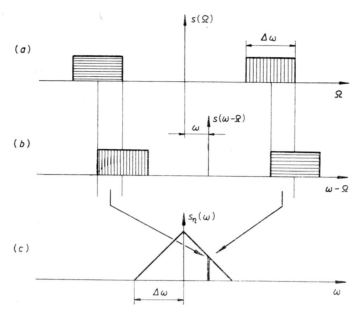

Figure 7-4 Mixing products of the output signal. *(a)* Power spectrum of input signal. *(b)* Power spectrum of input signal, inverted and shifted by ω. *(c)* Power spectrum of output signal.

Evidently, the output signal power spectrum comprises other components too. For a quadratic characteristic, the Fourier transform of Eq. (7-24) is

$$S_\eta(\omega) = \frac{1}{2\pi} \int_{-\infty}^{\infty} \Gamma_\eta(\tau) e^{-j\omega\tau}\, d\tau = b^2\sigma^4\delta(\omega) + 2b^2 \int_{-\infty}^{\infty} s(\Omega)s(\omega-\Omega)\, d\Omega \quad (7\text{-}28)$$

where Eqs. (3-64) and (3-77) have been used. The first term at the right-hand side is the DC component generated by the detector; a DC meter would read $b^2\sigma^4$.

7-3 SQUARE-LAW DETECTOR

Square-law detectors are frequently employed for the measurement of stochastic signals; the mean square value is directly proportional to the power. The same detector may be used for measuring signals with different shapes. Figure 7-5 shows the block diagram of the detector. A network with square-law transfer function is connected to a low-pass filter intended for attenuating the unwanted high-frequency components. In order to characterize the complete arrangement, the amplitude density function and the power spectrum of $\xi(t)$, $\eta(t)$, and $\zeta(t)$ are needed.

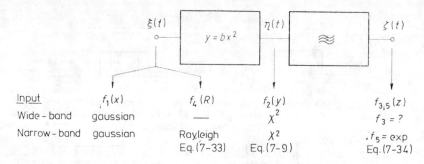

Figure 7-5 Block diagram of detector comprising low-pass filter, showing the distributions at the connection points.

Amplitude Density Function for a Wide-Band Input Signal

If the amplitude density function of the input signal $\xi(t)$ is $f_1(x)$, that of the squared signal $\eta(t)$ is $f_2(y)$, and $y = bx^2$, then according to Eq. (7-6),

$$f_2(y) = 2f_1(x) \left| \frac{dx}{dy} \right| = \frac{f_1(x)}{bx} = \frac{f_1[x(y)]}{\sqrt{by}} \tag{7-29}$$

In the case of a signal having a gaussian distribution, $f_2(y)$ is the density function of the χ^2 distribution with one degree of freedom [see Eq. (7-9)].

Unfortunately, there is no general procedure for the determination of $f_3(z)$—the density function of $\zeta(t)$. Based on Sec. 3-5, the only statement possible is that the distribution is somewhat normalized by the low-pass filter, so $f_3(z)$ is nearer to the normal distribution than $f_2(y)$.

Amplitude Density Function for an Input Signal with Small Relative Bandwidth

For this case, according to Fig. 2-10d and Eq. (2-131),

$$\xi(t) = \varrho(t) \cos\left[\omega_0 t + \vartheta(t)\right] \tag{7-30}$$

where $\varrho(t)$ is the stochastic time function of the envelope with Rayleigh distribution [see Eq. (2-139)], ω_0 is the band-center angular frequency, and $0 < \vartheta(t) < 2\pi$ is a random variable with even distribution. At the output of the network having the transfer function $y = bx^2$, by taking into account Eq. (7-30), we have

$$\eta(t) = \frac{b}{2} \varrho^2(t) + \frac{b}{2} \varrho^2(t) \cos\left[2\omega_0 t + 2\vartheta(t)\right] \tag{7-31}$$

If $\xi(t)$ has a normal distribution, the instantaneous values of $\eta(t)$ again follow the χ^2 distribution with one degree of freedom according to Eq. (7-9). However, the two terms of (7-31) may be separated by the low-pass filter. An ideal filter, which can be only approximated rather than realized, should pass the first term without attenuation and phase distortion but should show infinite attenuation for the second term (see Fig. 7-6c). Then

$$\zeta(t) = \frac{b}{2} \varrho^2(t) \tag{7-32}$$

the amplitude density function which may be calculated by the transformation given in Sec. 7-1. From Eq. (2-139), the amplitude density function of the input signal envelope is

$$f_4(R) = \frac{R}{\sigma^2} e^{-R^2/2\sigma^2} \qquad R \geq 0$$

the transfer characteristic is

$$z = \frac{b}{2} R^2 \tag{7-33}$$

and $f_5(z)$ pertaining to ζ, utilizing Eq. (7-5), is

$$f_5(z) = f_4(R) \left| \frac{dR}{dz} \right| = \frac{f_4(R)}{bR} = \frac{1}{b\sigma^2} e^{-z/b\sigma^2} \tag{7-34}$$

This is an exponential distribution. The relationships between the distributions considered are illustrated in Fig. 7-5.

Expected Values and Variances

Let the distribution of the input signal be normal, with zero expected value and standard deviation σ. The expected value of the detector output signal is, according to (2-122),

$$M(\eta) = \int_0^\infty y f_2(y) \, dy = b\sigma^2 \tag{7-35}$$

and from Eq. (2-121)

$$D^2(\eta) = M(\eta^2) - M^2(\eta)$$
$$= b^2 [M(\xi^4) - M^4(\xi)] = 2b^2\sigma^4 \tag{7-36}$$

Equation (7-35) gives the DC mean value, and Eq. (7-36) the mean square value of the fluctuation; the latter is present because of the AC components.

At the output of the low-pass filter, the DC mean value should be unchanged and the fluctuation, which is dependent on $f_3(z)$, should be less than that given by Eq. (7-36). Unfortunately, $f_3(z)$ is generally not known. However, for a narrow-band input signal, $f_5(z)$ can be interpreted by stating its expected value

$$M(\zeta) = \int_0^\infty z f_5(z)\, dz = b\sigma^2 \qquad (7\text{-}37)$$

Furthermore, the variance is

$$D^2(\zeta) = M(\zeta^2) - M^2(\zeta) = b^2\sigma^4 \qquad (7\text{-}38)$$

which is just half of that given by Eq. (7-36). Filtering out the double frequency term in Eq. (7-31), which does not include a DC component, has the effect of halving the mean square value of the fluctuation.

Power Spectrum

The power spectrum of $\eta(t)$ and $\zeta(t)$ may be determined utilizing the methods presented in Secs. 7-2 and 3-5. For simplicity, let the relative bandwidth $(\Delta\omega/\omega_0)$ of the input signal with normal distribution be small, and the power spectrum inside the pass-band constant (see Fig. 7-6) be

$$s_\xi(\omega) = A \qquad (7\text{-}39)$$

According to Eq. (3-70), the power of this signal is

$$\sigma^2 = D^2(\xi) = M(\xi^2) - 0 = \int_{-\infty}^\infty s_\xi(\omega)\, d\omega = 2A\Delta\omega \qquad (7\text{-}40)$$

Substituting this into Eq. (7-28), we have

$$s_\eta(\omega) = 4b^2 A^2 (\Delta\omega)^2 \delta(\omega)$$

$$+ \begin{cases} 4b^2 A^2 (\Delta\omega - |\omega|), & \text{if} \quad |\omega| \leq \Delta\omega \\ 2b^2 A^2 (\Delta\omega - \big||\omega| - 2\omega_0\big|), & \text{if} \quad 2\omega_0 - \Delta\omega < |\omega| < 2\omega_0 + \Delta\omega \\ 0 & \text{elsewhere} \end{cases} \qquad (7\text{-}41)$$

Figure 7-6b shows all the mixing products. The DC mean value—in accordance with Eq. (7-35)—is given by $4b^2 A^2 (\Delta\omega)^2$. All further components with frequencies differing from zero give rise to fluctuation. The fluctuation power is proportional to the areas bound by the triangles. If the low-pass filter with the characteristic shown in Fig. 7-6c is used to filter out the components symmetrically placed at $\pm 2\omega_0$ (see Fig. 7-6d) the fluctuation power is halved.

Our results can be extended to wide-band signals and signals with variable spectral densities. Equation (7-41) remains unchanged if f is written in place of all ω values and the two-sided power spectrum $s(f)$ is used instead of $s(\omega)$. The use of one-sided spectra is not feasible because the Dirac delta has to appear at the origin.

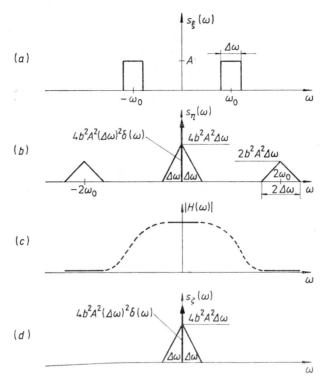

Figure 7-6 Power spectra of square-law detector. *(a)* Band-limited input signal. *(b)* Mixing products. *(c)* Response of ideal low-pass filter. *(d)* Power spectrum of low-pass filter.

7-4 LINEAR HALF-WAVE AND FULL-WAVE DETECTORS

A linear detector provides a DC mean value. In the case of a half-wave detector, this is the mean value of the positive instantaneous values, while in the case of a full-wave detector, this is the mean value of the (phase-inverted) negative and positive values. There is no unique correspondence between the DC mean value and the mean square value related to the power. Therefore, the linear detector produces a power-related result only for a signal with known amplitude density function. If this condition is met the use of a linear detector is warranted because of its simplicity.[6] Its characteristics will be treated later, but the more simple full-wave variant will first be considered for all parameters.

Amplitude Density Function

Let us again assume that the input signal $\xi(t)$ has normal distribution with zero expected value and standard deviation σ. The amplitude density function $f_2(y)$ of the output signal $\eta(t)$ is given by Eq. (7-11) for a full-wave detector and by Eq. (7-14) for a half-wave detector. Again, $f_3(z)$—related to the low-pass filter output signal—is not known.

Amplitude Density Function for an Input Signal with Small Relative Bandwidth

Utilizing Eq. (7-30), the output signal of the full-wave detector is

$$\eta(t) = a|\xi(t)| = a\varrho(t)\,|\cos[\omega_0 t + \vartheta(t)]| \tag{7-42}$$

and that of the half-wave detector is

$$\eta(t) = \begin{cases} a\varrho(t)\cos[\omega_0 t + \vartheta(t)] & \text{if} \quad -\dfrac{\pi}{2} < [\omega_0 t + \vartheta(t)] < \dfrac{\pi}{2} \\[2mm] 0 & \text{if} \quad \dfrac{\pi}{2} < [\omega_0 t + \vartheta(t)] < \dfrac{3\pi}{2} \end{cases} \tag{7-43}$$

The cosine term of Eq. (7-42) has a mean value of $2/\pi$ and that of (7-43) a mean value of $1/\pi$. If the conditions given in Sec. 7-3 are met, the output signal of the low-pass filter is

$$\zeta(t) = \frac{2}{\pi}\,a\varrho(t) \tag{7-44}$$

or

$$\zeta(t) = \frac{1}{\pi}\,a\varrho(t) \tag{7-45}$$

As ϱ has a Rayleigh distribution, ζ also has a Rayleigh distribution, so the density function at the output of the low-pass filter is

$$f_5(z) = \frac{\pi}{2a}\,\frac{\pi}{2a}\,\frac{z}{\sigma^2}\,e^{-\pi^2 z^2/8a^2\sigma^2} \tag{7-46}$$

For a half-wave rectifier, the output is

$$f_5(z) = \frac{\pi^2}{a^2}\,\frac{z}{\sigma^2}\,e^{-\pi^2 z^2/2a^2\sigma^2} \tag{7-47}$$

Expected Values and Variances

For the full-wave rectifier, Eq. (7-11) yields

$$M(\eta) = \int\limits_0^\infty y f_2(y)\, dy = \int\limits_0^\infty \frac{2y}{a\sigma \sqrt{2\pi}} e^{-y^2/2a^2\sigma^2}\, dy = \sqrt{\frac{2}{\pi}}\, a\sigma \qquad (7\text{-}48)$$

and for the half-wave rectifier, Eq. (7-14) yields

$$M(\eta) = \int\limits_0^\infty \frac{y}{a\sigma \sqrt{2\pi}} e^{-y^2/2a^2\sigma^2}\, dy = \frac{a\sigma}{\sqrt{2\pi}} \qquad (7\text{-}49)$$

From Eq. (2-125), the variance is

$$D^2(\eta) = a^2\sigma^2 \left(1 - \frac{2}{\pi}\right) \qquad (7\text{-}50)$$

and for the half-wave rectifier the variance is

$$D^2(\eta) = \frac{a^2\sigma^2}{2} \left(1 - \frac{1}{\pi}\right) \qquad (7\text{-}51)$$

The mean value at the low-pass filter output is

$$M(\zeta) = M(\eta) \qquad (7\text{-}52)$$

and the variance originating from the envelope fluctuation is

$$D^2(\zeta) = M(\zeta^2) - M^2(\zeta) = \frac{2}{\pi} a^2\sigma^2 \left(\frac{4}{\pi} - 1\right) \qquad (7\text{-}53)$$

and for the half-wave rectifier is

$$D^2(\zeta) = \frac{1}{2\pi} a^2\sigma^2 \left(\frac{4}{\pi} - 1\right) \qquad (7\text{-}54)$$

The filtering out of the high-frequency components will, in this case, also substantially reduce the fluctuation.

Power Spectrum

In order to determine the power spectrum of $\eta(t)$, a knowledge of the autocorrelation function of $\Gamma_\eta(\tau)$ is necessary. Omitting the details of derivation,[8, 20] for a full-wave detector, we have

$$\Gamma_\eta(\tau) \cong a^2 \left[\frac{2}{\pi} \sigma^2 + \frac{1}{\pi} \frac{\Gamma_\xi^2(\tau)}{\sigma^2}\right] \qquad (7\text{-}55)$$

and for a half-wave detector

$$\Gamma_\eta(\tau) \cong a^2 \left[\frac{1}{2\pi} \sigma^2 + \frac{1}{4} \Gamma_\xi(\tau) + \frac{1}{4\pi} \frac{\Gamma_\xi^2(\tau)}{\sigma^2}\right] \qquad (7\text{-}56)$$

Expression (7-55) is very similar to (7-24). Furthermore, Eq. (7-56) even comprises the first power of the autocorrelation function $\Gamma_\xi(\tau)$ pertaining to the input signal. The Fourier transform of the previous relationships yields the respective power spectra

$$s_\eta(\omega) \cong \frac{2}{\pi} a^2 \sigma^2 \delta(\omega) + \frac{a^2}{\pi\sigma^2} \int_{-\infty}^{\infty} s(\Omega) s(\omega-\Omega) \, d\Omega \qquad (7\text{-}57)$$

and

$$s_\eta(\omega) \cong \frac{a^2\sigma^2}{2\pi} \delta(\omega) + \frac{a^2}{4} s_\xi(\omega) + \frac{a^2}{4\pi\sigma^2} \int_{-\infty}^{\infty} s(\Omega) s(\omega-\Omega) \, d\Omega \qquad (7\text{-}58)$$

These are shown in Fig. 7-7.

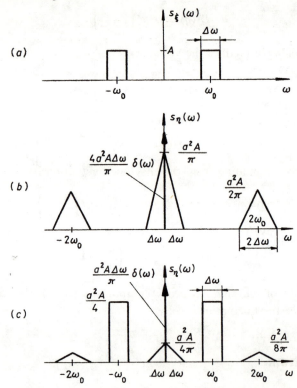

Figure 7-7 (*a*) Band-limited input signal. (*b*) Approximate output power spectrum for full-wave linear detector. (*c*) Approximate output power spectrum for half-wave linear detector.

Due to the asymmetry of the half-wave detector characteristic, the output will comprise the input signal spectrum (base-band) too, and this increases the overall fluctuation.[22] This is also clearly seen by comparison of the relative fluctuations calculated from Eqs. (7-50) and (7-48), or (7-51) and (7-49), respectively:

$$\frac{D(\eta)}{M(\eta)} \text{ (half-wave): } \frac{D(\eta)}{M(\eta)} \text{ (full-wave)} = \sqrt{2\frac{\pi-1}{\pi-2}} = 1.94 \qquad (7\text{-}59)$$

The filtering out of the base-band is much more difficult than that of the sum components, which have much higher frequencies, and is not at all possible in the case of $\Delta\omega > 2/3\omega_0$.

7-5 PEAK DETECTOR

For the analysis of the square-law and linear detectors, the general variables ξ, η, x, and y have been used as both the input and output signals may be either voltage or current. However, the usual circuit of the peak detector shown in Fig. 7-8a assumes a voltage generator drive, and the output signal is usually the voltage appearing across the capacitor. The diode characteristic is approximated by two straight lines, and its series resistance is embedded in r.

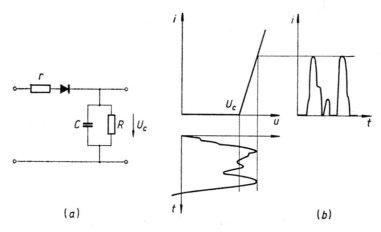

Figure 7-8 *(a)* Circuit diagram of peak detector. *(b)* Graphical determination of the current flowing through the diode.

The peak detector characteristic is shown in Fig. 7-8b. The break point is shifted and its location depends on U_C. Let us assume that $C=\infty$, so U_C is constant. In the stationary case, the charges fed into and taken out of the capacitor have to be equal. The equation

$$\frac{U_C}{R} = I = M(i) = \frac{1}{r}\int_{U_C}^{\infty} (u - U_C)f(u)\,du \qquad (7\text{-}60)$$

then holds.[19] After a slight rearrangement, Eq. (7-60) may be written as

$$\frac{1}{U_C} \int_{U_C}^{\infty} u f(u)\, du - \int_{U_C}^{\infty} f(u)\, du = \frac{r}{R} \tag{7-61}$$

If there is a gaussian distribution with a density $f(u)$ and standard deviation σ_u, Eq. (7-61) takes the form

$$\frac{1}{U_C} \frac{\sigma_u}{\sqrt{2\pi}} e^{-U_C^2/2\sigma_u^2} - \left[1 - \Phi\left(\frac{U_C}{\sigma_u}\right)\right] = \frac{r}{R} \tag{7-62}$$

where $\Phi(U_C/\sigma_u)$ is the distribution function defined in Eq. (2-65). The relationship between U_C/σ_u and r/R is shown by the upper curve of Fig. 7-9.

This result, for a wide-band input signal, is valid as long as $\omega_l > 1/RC$ where ω_l is the lower limit of the input signal frequency range. On the other hand, two cases are possible for an input signal with a relatively narrow bandwidth.

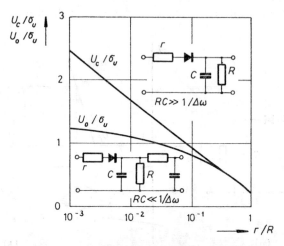

Figure 7-9 Comparison of peak detectors with high time constant and low time constant, the latter comprises an additional low-pass filter.

1. $RC \gg 1/\Delta\omega$. Neither the carrier ω_0 nor the envelope comprising the highest frequency component of $\Delta\omega/2$ will appear at the output. As the output signal is pure DC voltage Eq. (7-62) may be used.
2. $\Delta\omega \ll 1/RC \ll \omega_0$. The envelope appears at the output which would just be at a tangent to the half sine-wave pulse peaks shown in Fig. P2-3c if series resistance r were zero. The loss due to the finite r/R ratio can again be calculated using Eq. (7-61), but the amplitude density function of the sine wave

[see Eq. (1-15)] has to be substituted in place of $f(u)$. Thus

$$\frac{1}{\pi}\left[\sqrt{\left(\frac{U_{0e}}{U_{ce}}\right)^2 - 1} + \arcsin\left(\frac{U_{ce}}{U_{0e}}\right)\right] - \frac{1}{2} = \frac{r}{R} \tag{7-63}$$

where U_{0e} is the instantaneous value of the envelope and U_{ce} is the instantaneous value of the capacitor voltage. Taking into account the relationship for the mean value of the envelope given by Eq. (2-141) that

$$M(\varrho) = \sqrt{\frac{\pi}{2}}\,\sigma_u \cong 1.25\sigma_u$$

the output DC mean voltage, depending on r/R, will be $U_0 = 1.25(U_{ce}/U_{0e})\sigma_u$ (lower curve in Fig. 7-9). This is smaller than or, at most, equal to the value appearing across the RC network without additional filtering.

It can be seen from Fig. 7-9 that the output DC voltage is not an unambiguous function of σ_u. If, for example, the conditions applying to the RC time constant are not met, then U_0/σ_u will fall between the two curves. Therefore, measuring devices employing peak detectors, such as simple electronic voltmeters, may only be used with extreme care after previous calibration with a signal of appropriate frequency range, bandwidth, and amplitude density function.

7-6 NONLINEARITY OF THE STAGES PRECEDING THE DETECTOR

The realizable detectors treated in Sec. 7-7 generally require high input levels for proper functioning, so a high-level driving amplifier is required. The level requirement is further increased by the type of stochastic signal amplitude density function. At normal distribution, infinitely high instantaneous values may appear theoretically, although the probability of instantaneous values higher than $3\sigma_n$ appearing is as low as 0.3 percent.

The nonlinearity of the amplifier transfer function has an effect on both the amplitude density function and the power spectrum. The calculations may be carried out utilizing the methods considered in Secs. 7-1 and 7-2. If the equation for the transfer characteristic is in the form $y = g(x)$ [see Eq. (7-1)], the modification of the amplitude density function may be calculated from Eq. (7-5). For low distortions, the change in the power spectrum is less significant and its calculation is, in general, extremely difficult as the nonlinearity of the characteristic may only be taken into account by the transformation method mentioned briefly in Sec. 7-2.

On the other hand, the combined behavior of an amplifier with sharp limiting characteristic and a full-wave linear or square-law detector can easily be determined. Let the equation for the detector characteristic, according to Eq. (7-10), be

$$y = a|x| \qquad (7\text{-}64)$$

and the characteristic of the amplifier be

$$x = \begin{cases} Kv & \text{if} \quad -x_0 < x < x_0 \\ \pm x_0 & \text{if} \quad x < -x_0 \quad \text{or} \quad x > x_0 \end{cases} \qquad (7\text{-}65)$$

where K is the gain and v is the amplifier input signal.

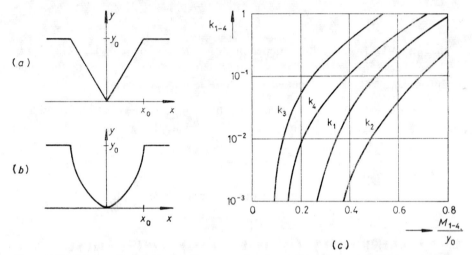

Figure 7-10 *(a)* Saturated full-wave linear characteristic. *(b)* Saturated full-wave square-law characteristic. *(c)* Correction factors: k_1—wide-band input signal, linear detection; k_2—narrow-band input signal, linear detection; k_3—wide-band input signal, square-law detection; k_4—narrow-band input signal, square-law detection. M_1 to M_4 are measured mean values of the output signal. *(Courtesy of American Institute of Physics and the author.[7])*

The overall characteristic is shown in Fig. 7-10a. The mean value of the output signal can be calculated from

$$M(\eta) = 2 \int_0^{y_0} yf(y)\,dy + 2y_0 \int_{y_0}^{\infty} f(y)\,dy \qquad y_0 = ax_0 \qquad (7\text{-}66)$$

as everywhere y_0 has to be written with a frequency of $f(y)$, instead of the values $y > y_0$ which do not occur. In the case of normal distribution

$$M(\eta) = \sigma_n \sqrt{\frac{2}{\pi}}(1 - e^{-y_0^2/2\sigma_n^2}) + y_0\left[1 - \Phi\left(\frac{y_0}{\sigma_n\sqrt{2}}\right)\right] \qquad (7\text{-}67)$$

which tends to $\sigma_n\sqrt{2/\pi}$ if $y_0 \to \infty$. The correction factor

$$\frac{\sigma_n\sqrt{2\pi}}{M(\eta)} - 1 = k_1\left[\frac{M(\eta)}{y_0}\right] \qquad (7\text{-}68)$$

is shown in Fig. 7-10c. Similar curves also apply for an input signal with Rayleigh distribution or for a square-law detector (correction factors k_2, k_3, k_4). The error due to amplifier saturation is most significant for square-law detection of a wide-band signal (k_3); the weight of extreme instantaneous values is the highest in this case.

7-7 REALIZATION OF DETECTORS

It is not possible to realize a detector with a characteristic which would be equal to the ideal functions (7-7), (7-10), and (7-12) in an arbitrarily wide range. The degree of approximation depends on the difference between the ideal characteristic and the characteristic achieved by actual devices.

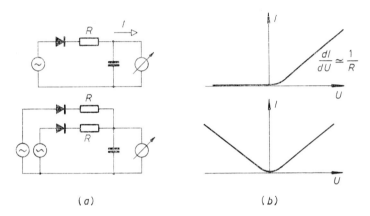

Figure 7-11 *(a)* Circuit diagrams of half-wave and full-wave linear detectors. *(b)* Characteristics due to nonideal diodes.

Figure 7-11 shows the simple forms of a half-wave and a full-wave linear detector and their characteristics. For satisfactory operation, an input voltage of several volts is needed because the bend at the origin is not sharp enough. The ratio of the curvature and the slope

$$\frac{d^2 I/dU^2}{dI/dU} = \frac{1}{U_T} = \frac{q}{kT} \tag{7-69}$$

is the same for all p–n junctions, and depends only on the temperature.

The ideal piecewise linear characteristic may be much better approximated by diodes combined with operational amplifiers. The half-wave rectifier shown in Fig. 7-12a operates as follows: for an input signal with negative instantaneous values U_0 is positive, and due to the insertion of the feedback path D_1–R–R, the gain is unity. Should the output signal become negative, the gain is reduced to practically zero by the feedback loop closed through diode D_2.[25]

There are several methods of realizing a full-wave arrangement.[2] One is to connect in parallel to the inverting network with the piecewise linear characteristic shown in Fig. 7-12a another, equally piecewise linear, but noninverting, network.[25] A further variant is shown in Fig. 7-12b. The diode bridge is again in the feedback path, so the nonlinearity of the diode characteristic is decreased by the loop gain. It is important to note that here the output quantity is current and the load resistance may not be grounded. However, this disadvantage may be eliminated using a further amplifier with symmetrical input and asymmetrical output.[13]

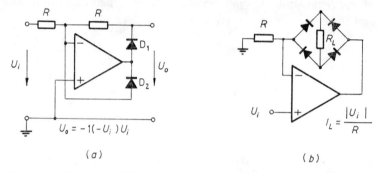

Figure 7-12 Detector utilizing an operational amplifier for the improvement of detection characteristics. *(a)* Half-wave detector, *(b)* Full-wave detector.

The unambiguous measurement of the mean square value in the case of unknown signals or signals with varying amplitude density function is only possible by using more complicated detectors which approximate the square-law characteristic. Their operation is based on one of the following parameters:[24]

1. The heating effect of the current
2. Nonlinear characteristic of an electronic device
3. Bilinear characteristic of an electronic device
4. Piecewise linear approximation of the square-law characteristic

Advantages and disadvantages of these parameters will be compared, without aiming at complete coverage, with the aid of the following examples.

1. The direct measurement of the mean square value of the current is possible using a filament lamp, the filament of a vacuum tube, a thermocouple, and a thermistor,[4, 5, 21] or an integrated circuit including a temperature sensor and a heating element.[28, 32] Absolute measurements are also possible utilizing these devices but they are hampered by their slow response, sensitivity to overload, and temperature dependence.

2. The current–voltage characteristic of electronic devices may be expanded into the form

$$I = I_0 + a_1 U + a_2 U^2 + \ldots + a_n U^n \qquad (7\text{-}70)$$

The current sum of two push–pull-driven devices (Fig. 7-13) is

$$\Sigma I = 2I_0 + 2a_2 U^2 + 2a_4 U^4 + \ldots \qquad (7\text{-}71)$$

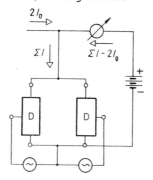

Figure 7-13 Push–pull-driven square-law detector, realized with nonlinear devices.

The quiescent current $2I_0$ may be eliminated by compensation; the fourth- and higher-order terms are generally negligible. Best results are achieved utilizing devices with inherently approximate square-law characteristics, such as JFETs or MOS field-effect transistors,[15, 23] or tunnel diodes.[29] However, temperature dependence may also be troublesome. Typical nonlinearity is given by the exponential characteristic of bipolar devices. This may also be used to achieve a square-law characteristic, according to the equation[24]

$$U_0 = U_2 e^{2\log U_{\text{in}}/U_1} = bU_{\text{in}}^2 \qquad (7\text{-}72)$$

where U_1 and U_2 are the constants of the logarithmic and exponential function generators. However, it is important to note that these function generators require DC voltages, so the signal to be measured has first to be rectified and filtered. This way independence of the amplitude density function of the signal to be measured may be lost.

3. Presently, the bilinear multiplying characteristic can most frequently be realized using an integrated circuit based on the skilful utilization of the p–n junction exponential characteristic.[9, 10] The output signal of this circuit —in a limited frequency range—is proportional to the product of the input

signal instantaneous values. This solution is thus free from the restriction applicable to the circuit described by Eq. (7-72).

4. Figure 7-14 shows the circuit diagram and the characteristic of a square-law detector with piecewise linear approximation.[14, 31] By increasing the voltage of the signal to be rectified, more and more diode–resistance networks are connected in parallel, so the slope of the current–voltage characteristic is increased stepwise. The piecewise linear characteristic, made up of straight-line sections, approximates a parabola. The approximation is improved by more closely placed break points. Break point locations are determined by a given absolute or relative error limit.[17]

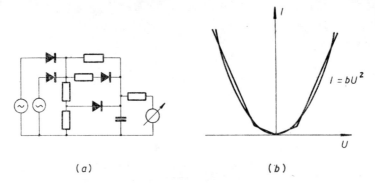

$$I = bU^2$$

(a)   (b)

Figure 7-14 Detector with piecewise linear characteristic, with two break points at each parabola branch. *(a)* Circuit diagram. *(b)* Characteristic.

When a periodic or stochastic signal is detected, the input signal sweeps the sections of the characteristic which have a positive or negative error $h(u)$. According to Fig. 7-15a, the overall dynamic error h_d is

$$h_d = \int_{-\infty}^{\infty} h(u)f(u)\,du \tag{7-73}$$

This error is much smaller than the static error.[1, 27] The distance between break points may be significantly increased thus increasing the drive capability or requiring a characteristic with less break points for a given drive range (see Fig. 7-15c).

Here again, the temperature dependence and finite curvature of the diode characteristic result in errors that can significantly be reduced using operational amplifiers.[16, 18]

It should be noted for completeness that there are detectors both in piecewise linear and in bipolar integrated forms suitable for generating the square root of the sum of two or more quantities squared.[11, 12, 26]

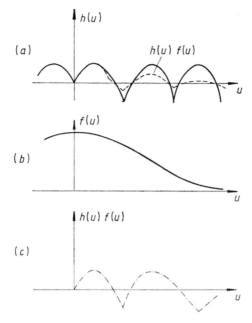

Figure 7-15 Detector with piecewise linear characteristic. *(a)* Steady state error. *(b)* Input signal density function. *(c)* $h(u)f(u)$ in case of increased distance between break points.

PROBLEMS

7-1 White noise, passed by a low-pass filter, is detected by a square-law detector. What is the power spectrum of the output signal?

Answer: The power spectra of the input and output signals are shown by Figs. P7-1a and P7-1b, respectively. According to Eqs. (7-39), (7-40), and (7-41) the output signal is made up of the DC component $4b^2 A^2 (\Delta\omega)^2 \delta(\omega)$ and the fluctuation component $2b^2 \int_{-\infty}^{\infty} s(\Omega)s(\omega-\Omega)d\Omega$. The convolution of a spectrum of width $2\Delta\omega$ is an isosceles triangle of width $2(2\Delta\omega)$. The height of this triangle is

$$2b^2 \int_{-2\Delta\omega}^{0} A^2 \, d\Omega = 4b^2 A^2 \, \Delta\omega$$

Comparing this with Fig. 7-6, it is noted that all fluctuation components are now clustered in a band of $\pm 2\Delta\omega$ around the origin.

7-2 Consider the peak detector shown in Fig. 7-8a. Calculate the ratio r/R for which the detector supplies the same DC voltage for a sine wave and a noise of normal distribution having the same mean squared value.

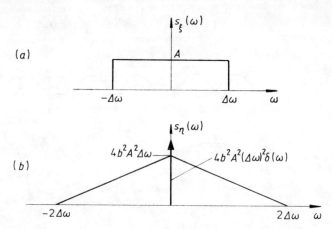

Figure P7-1 *(a)* Power spectrum of low-pass filtered white noise.
(b) Power spectrum of square-law detected white noise.

Answer: This requires that Eqs. (7-62) and (7-63) be equal. Taking into account that in this case $\sigma_{u\,noise} = \sigma_{u\,sin} = U_0/\sqrt{2}$, the following transcendental equation has to be solved:

$$\frac{1}{U_C}\frac{\sigma}{\sqrt{2\pi}}e^{-U_C^2/2\sigma^2} + \Phi\left(\frac{U_C}{\sigma}\right) - 1 = \frac{1}{\pi}\left(\sqrt{\frac{2\sigma^2}{U_C^2}-1} + \arcsin\frac{U_C}{\sigma\sqrt{2}}\right) - \frac{1}{2}$$

From this equation $U_C/\sigma \cong 0.83$ and $r/R \cong 0.137$.

REFERENCES

[1] Ambrózy, A.: "Die dynamischen Fehler polygonal angenäherter quadratischer Detektoren", *ATM*, vol. 311, no. 12, pp. 165—170, 1961.

[2] Antoniou, A.: "Design of Precision Rectifiers Using Operational Amplifiers", *Proc. IEE.*, vol. 121, no. 10, pp. 1041—1044, 1974.

[3] Bennett, W. R. and S. O. Rice: "Note on Methods of Computing Modulation Products", *Phil. Mag.*, vol. 18, no. 9, pp. 422—424, 1934.

[4] Bozic, S. M.: "Mean Square Voltmeter", *J. Sci. Instr.*, vol. 39, no. 5, pp. 210—213, 1962.

[5] — and J. R. Whitbread: "Thermistor Power Meter", *Electr. Eng.*, vol. 39, no. 10, pp. 637—639, 1967.

[6] Cohn, C. E.: "Linear Detectors for Noise Measurements", *Rev. Sci. Instr.*, vol. 35, no. 5, pp. 638—639, 1964.

[7] —: "Errors in Noise Measurement due to the Finite Amplitude Range of the Measuring Instrument", *Rev. Sci. Instr.*, vol. 35, no. 6, pp. 701—703, 1964.

[8] Davenport, W. B. and W. L. Root: *An Introduction to the Theory of Random Signals and Noise*, McGraw-Hill, New York, 1958.

[9] Gilbert, B.: "A New Wideband Amplifier Technique", *IEEE J.*, vol. SC-3, no. 4, pp. 353—365, 1968.

[10] —: "A High Performance Monolithic Multiplier Using Active Feedback", *IEEE J.*, vol. SC-9, no. 6, pp. 364—373, 1974.

[11] —: "High-Accuracy Vector-Difference and Vector-Sum Circuits", *Electr. Lett.*, vol. 12, no. 11, pp. 293—294, 1976.

[12] —: "General Technique for n-Dimensional Vector Summation of Bipolar Signals", *Electr. Lett.*, vol. 12, no. 19, pp. 504—505, 1976.

[13] Graeme, J. G.: *Designing with Operational Amplifiers*, McGraw-Hill, New York, 1977.

[14] Hansen, J. A.: "RMS Rectifiers", *Brüel–Kjaer Tech. Rev.*, vol. 19, no. 2, pp. 3—18, 1972.

[15] Házman, I.: "Four Quadrant Multiplier Using MOSFET Differential Amplifiers", *Electr. Lett.*, vol. 8, no. 3, pp. 63—65, 1972.

[16] Herpy, M.: *Analog Integrated Circuits*, John Wiley, New York, 1979.

[17] Marjanovic, S.: "High Precision $y=x^2$ Generator", *Electr. Lett.*, vol. 12, no. 7, pp. 165—166, 1976.

[18] Ochs, G. and P. Richman: "Curve Fitter for RMS Measurement", *Electronics*, vol. 42, no. 20, pp. 98—101, 1969.

[19] Peterson, A. P. G.: "Response of Peak Voltmeters to Random Noise", *Gen. Radio Exp.*, vol. 31, no. 7, pp. 3—8, 1956.

[20] Rice, S. O.: "Mathematical Analysis of Random Noise", *BSTJ*, vol. 23, no. 3, pp. 282—333, 1944; vol. 24, no. 1, pp. 46—157, 1945.

[21] Richman, P. L.: "A New Wideband True RMS to DC Converter", *IEEE Trans.*, vol. IM-16, no. 2, pp. 129—134, 1967.

[22] Rubin, M. D. "Comparison of Signal and Noise in Full Wave and Half Wave Rectifiers", *IEEE Trans.*, vol. IT-8, no. 5, pp. 379—380, 1962.

[23] Sevin, L. J.: *Field-Effect Transistors*, McGraw-Hill, New York, 1965.

[24] Sheingold, D. H.: *Nonlinear Circuits Handbook*, 2nd ed., Analog Devices, Norwood, 1976.

[25] Smith, J. I.: *Modern Operational Circuit Design*, John Wiley, New York, 1971.

[26] Stern, T. E. and R. M. Lerner: "A Circuit for the Square Root of the Sum of the Squares", *Proc. IEEE*, vol. 51, no. 4, pp. 593—596, 1963.

[27] Száraz, G.: "Accuracy Measurement of Root-Mean-Square-Measuring Instruments" (Effektív értéket mérő műszerek pontosságának mérése), *Híradástechnika*, vol. 18, no. 12, pp. 383—388, 1967, (in Hungarian).

[28] Székely, V.: "New Type of Thermal Function IC: The Four-Quadrant Multiplier", *Electr. Lett.*, vol. 12, no. 15, pp. 372—373, 1976.

[29] Tarnay, K.: "True RMS Measurement with Tunnel Diodes", *Proc. IEEE*, vol. 50, no. 10, p. 2124, 1962.

[30] van der Ziel, A.: *Noise*, Prentice Hall, Englewood Cliffs, 1954.

[31] Wahrman, C. G.: "A True RMS Instrument", *Brüel–Kjaer Tech. Rev.*, vol. 5, no. 3, pp. 9—21, 1958.

[32] *Electronics*, vol. 48, no. 21, p. 94, 1975.

EIGHT

NOISE MEASUREMENTS

The noise equivalent circuits of active electronic devices may not always be based on theory, and the validity of seemingly simple noise models must be checked by measurements.[34,35] Also, the avalanche, flicker, and burst noise may only be determined empirically. A better understanding of these components may result in devices of higher quality due to a reduction in noise. In some cases, noise measurement results allow the determination of active device parameters which otherwise could only be measured inaccurately and with difficulty (e.g., the resistance $r_{bb'}$ of a bipolar transistor, lead resistances of a field-effect transistor). Before investigating measurement methods, the theoretical limitations of measurement accuracy will be surveyed.

8-1 TIME REQUIREMENT

Figure 8-1 shows the general block diagram of a noise measurement system. The band-pass filter of bandwidth Δf may be realized by analog or digital methods; it may not be separately included because the coupling elements and load reactances of the amplifiers may limit the transmitted frequency band from both sides. The detector may have linear or square-law characteristics. For measurement,

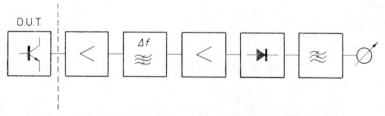

Figure 8-1 Simplified block diagram of noise measurement.
D.U.T.: device under test.

the DC component of the detected signal is of importance. However, onto this component is superimposed the fluctuation of the mean square value $D^2(\eta)$ (see Chapter 7). For this reason, the filtering out of the fluctuation components in the detected signal is essential; this is accomplished using an analog or digital integrator (low-pass filter).

Due to the limited bandwidth, the noise measurement has a finite time requirement.[16] This is explained by the fact that according to the sampling theorem, given a signal of bandwidth Δf, the highest number of independent samples taken from this signal may be $2\Delta f$ per second. This means that during a time interval T, the highest number of samples may be $2\Delta fT$ (see Fig. 8-2).

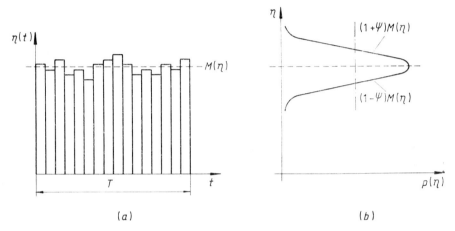

(a) (b)

Figure 8-2 Independent samples originating from a band-limited signal.
(a) Time diagram. *(b)* Amplitude distribution.

In the case of square-law detection, the sample amplitudes have a χ^2-distribution according to Eq. (7-9). Assume that the sum of $K=2\pi fT$ is the number of sample amplitudes stored by an ideal integrator. Utilizing Eqs. (2-119) and (7-35), the expected value of the random variable determined by summation of the square-law detected signals is

$$M(\eta_1+\eta_2+\ldots+\eta_K) = bM(\xi_1^2+\xi_2^2+\ldots+\xi_K^2) = Kb\sigma^2 \qquad (8\text{-}1)$$

and according to Eq. (2-121), the variance is

$$D^2(\eta_1+\eta_2+\ldots+\eta_K) = b^2 D^2(\xi_1^2+\xi_2^2+\ldots+\xi_K^2) = 2Kb^2\sigma^4 \qquad (8\text{-}2)$$

Utilizing Eqs. (2-128) and (7-48), the same parameters for the full-wave linear detector are

$$M(\eta_1+\eta_2+\ldots+\eta_K) = KM(\eta) = \sqrt{\frac{2}{\pi}}\, aK\sigma \qquad (8\text{-}3)$$

and from Eq. (2-129)

$$D^2(\eta_1+\eta_2+...+\eta_K) = KD^2(\eta) = \left(1-\frac{2}{\pi}\right)a^2 K\sigma^2 \tag{8-4}$$

We now introduce the ratio

$$\psi = \frac{D\left(\sum_1^K \eta_i\right)}{M\left(\sum_1^K \eta_i\right)} \tag{8-5}$$

for describing the relative standard deviation. In the case of a square-law detector, we have from Eqs. (8-1) and (8-2)

$$\boxed{\psi_s = \sqrt{\frac{2}{K}} = \frac{1}{\sqrt{\Delta f T}}} \tag{8-6}$$

and for a linear full-wave detector, we have from Eqs. (8-3) and (8-4)

$$\boxed{\psi_1 = \sqrt{\frac{\pi}{2}-1}\,\frac{1}{\sqrt{K}} = \frac{0.53}{\sqrt{\Delta f T}}} \tag{8-7}$$

With an increasing number of terms, the sum distributions $\sum_1^K \eta_i$ tend to a normal distribution. For $K \geq 30$, the error introduced by the approximation by normal distribution will be negligible. Then it may be stated according to Eqs. (2-67) and (8-5) that the probabilities of $\sum_1^K \eta_i$ falling between the limits

$$(1\pm m\psi)\,M\left(\sum_1^K \eta_i\right) \tag{8-8}$$

will be 63.8, 95.4 and 99.7 percent for $m=1, 2$, and 3, respectively. In the following, let us consider the product

$$m\psi = m(p)\psi = h \tag{8-9}$$

which yields the relative measurement error where $m(p)$ depends on the confidence of not exceeding the error limit (e.g., a confidence of $p=95$ percent means that the actual error will exceed $|m\psi|$ once in every 20 cases). Comparing Eqs. (8-6), (8-7) and (8-9), the minimum time requirement for measurement of error h and confidence p, with a predetector bandwidth of Δf, will be for a square-wave detector

$$T_{\min} = \frac{m^2(p)}{h^2\,\Delta f} \tag{8-10}$$

and for a linear double-wave detector

$$T_{\min} = 0.285\frac{m^2(p)}{h^2\,\Delta f} \tag{8-11}$$

Thus the measurement time may not be reduced arbitrarily.

The output signal of the low-pass filter shown in Fig. 8-1 has a similar error.[30, 32] This error is introduced by the fluctuation components shown in Figs. 7-6 and 7-7. Let the low-pass filter be a simple RC-network having a transfer function of

$$H(\omega) = \frac{1}{1+j\omega RC} = \frac{1}{1+j\omega/\omega_a} \tag{8-12}$$

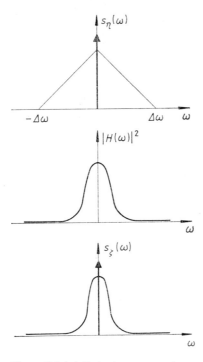

Figure 8-3 *(a)* Output power spectrum of square-law detector. *(b)* Frequency response of low-pass filter. *(c)* Power spectrum at the output of the filter, Eq. (8-13).

For good filtering, $\omega_a \ll \Delta\omega$ (see Fig. 8-3b). According to Eqs. (7-41) and (3-104), the power spectrum at the low-pass filter output, in the case of square-law detection, is

$$s_\zeta(\omega) = \frac{1}{1+(\omega/\omega_a)^2}[4b^2A^2\Delta\omega^2\delta(\omega)+4b^2A^2(\Delta\omega-|\omega|)] \tag{8-13}$$

In the expression

$$\int_{-\Delta\omega}^{\Delta\omega} s_\zeta(\omega)\,d\omega = 4b^2A^2\Delta\omega^2+4b^2A^2\int_{-\Delta\omega}^{\Delta\omega}\frac{\Delta\omega-|\omega|}{1+(\omega/\omega_a)^2}\,d\omega \tag{8-14}$$

the first right-hand side term is the mean square DC value

$$M^2(\zeta) = 4b^2 A^2 \Delta\omega^2 \tag{8-15}$$

and the second term is the squared overall fluctuation

$$D^2(\zeta) = 4b^2 A^2 \omega_a \left\{ \pi\Delta\omega - \omega_a \ln\left[1 + \left(\frac{\Delta\omega}{\omega_a}\right)^2\right]\right\} \tag{8-16}$$

The ratio of Eqs. (8-16) and (8-15)

$$v_s^2 = \frac{D^2(\zeta)}{M^2(\zeta)} \approx \frac{\omega_a}{\Delta\omega}\left[\pi - \frac{\omega_a}{\Delta\omega}\left(2\ln\frac{\Delta\omega}{\omega_a}\right)\right] \tag{8-17}$$

is the relative fluctuation squared when $\Delta\omega/\omega_a \gg 1$. Here the second term goes to zero if $\omega_a/\Delta\omega \to 0$, i.e. the filter time constant is high. After the decay of the switch-on transient (see Sec. 8-2), the relative fluctuation is

$$\boxed{v_s \cong \sqrt{\pi\frac{\omega_a}{\Delta\omega}} = \frac{1}{\sqrt{2\Delta f \tau_a}}} \tag{8-18}$$

where the time constant $\tau_a = RC = 1/\omega_a$ has been introduced. The meter shown in Fig. 8-1, which is assumed to have no inertia, will fluctuate around a mean value $M(\xi)$ with a standard deviation of $v_s M(\zeta)$.

Comparison of Figs. 7-6 and 7-7 shows that the fluctuation of a meter connected to a linear full-wave detector may be similarly calculated. The end result is

$$\boxed{v_l \cong \sqrt{\frac{\pi}{4}\frac{\omega_a}{\Delta\omega}} = \frac{1}{2}\frac{1}{\sqrt{2\Delta f \tau_a}}} \tag{8-19}$$

which differs only in the multiplying factor $1/2$ from Eq. (8-18). Note that the spectra shown in Figs. 7-7b and 7-7c are based on approximations.

If the low-pass filter bandwidth ω_a is much less than $\Delta\omega$, i.e. $\tau_a \gg 1/\Delta\omega$, then the low-pass filter *remembers* many periods of the detected signal. Its output signal may thus be regarded as a distribution of a multiterm sum, and the meter fluctuation has normal distribution. A small relative error may be achieved using a low-pass filter with a high time constant.

8-2 ANALOG INTEGRATION

The store function treated in Sec. 8-1 is realized by an ideal integrator described by the function

$$Y = \int_0^T \eta(t)\,dt \tag{8-20}$$

According to Eqs. (8-10) and (8-11), long integrating time is needed to achieve a measurement of high accuracy, and the end result may only be read after this time has elapsed. A further drawback is because of the increase of Y with T; this may impose requirements on the finite drive range of realizable integrators.

Figure 8-4 shows the principle of an analog integrator realized using a current generator and a capacitor. The finite leakage of the capacitor results in a non-ideal operation of this integrator. After the end of the integration time T—which has to be indicated by the measuring instrument—the stored value must be read as soon as possible. The application of such an analog integrator has thus severe drawbacks.

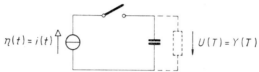

Figure 8-4 Analog integrator.

Continuous measurement is possible using the RC low-pass filter shown in Fig. 8-5a. Both charge and discharge may take place through resistance R and the load resistance of the detector, so in the stationary state, the capacitor voltage fluctuates around the mean value. However, the stationary state is reached in a finite time: a time interval of $t=RC$ allows an approximation of 63.2 percent, and a time interval of $t=2RC$ an approximation of 86.5 percent if the output voltage was zero before switch-on (see Fig. 8-5b). How many time constants should the time interval be from switch-on to read-off?

A higher time interval results in a better approximation of the mean value, but the fluctuation will remain constant determined by the time constant τ_a. At the same time, the quantity $K=2\Delta fT$ will increase, and its relative fluctuation will decrease; more information is supplied by the independent samples gathered during time T than utilized by the RC low-pass filter.

If the error of the maximum quantity of information that may be gathered in the interval 0 to T is equal to the fluctuation in the stationary state, i.e. $\Psi_s=\nu_s$ then comparison of Eqs. (8-6) and (8-18) yields

$$\frac{1}{\Delta fT} = \frac{1}{2\Delta f\tau_a} \tag{8-21}$$

i.e. $\tau_a=T/2$. A similar result is derived for linear detection too, by comparing Eqs. (8-7) and (8-19). Thus

$$\frac{0.285}{\Delta fT} \cong \frac{1}{4}\frac{1}{2\Delta f\tau_a} \tag{8-22}$$

i.e. $\tau_a \cong 0.88\,(T/2) \approx T/2$.[1,2] This means that it is not worthwhile waiting longer than time $T=2\tau_a$ to read the output signal at the RC low-pass filter output.[14]

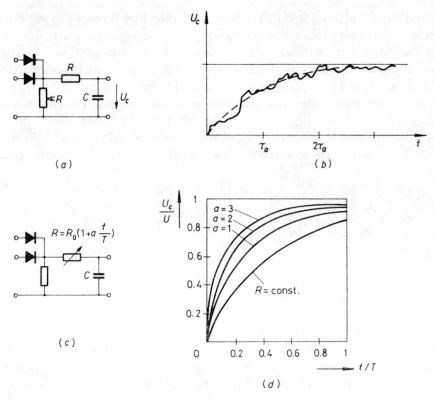

Figure 8-5 *(a)* Low-pass filter with constant time constant. *(b)* Time function of output signal. *(c)* Low-pass filter with varying time constant. *(d)* Time function of the output signal expected value.

Unfortunately, this time is insufficient for the switch-on transient to disappear. At the time instant $t=2\tau_a$, the residual error will be $e^{-2}=0.135=13.5$ percent. The solution is an RC network with varying time constant (see Fig. 8-5c). At the beginning, the resistance and thus the time constant are low but then gradually—possibly in a stepwise manner—they increase up to the value determined by the fluctuation allowed in the stationary state. Figure 8-5d shows the switch-on transients for a resistance–time function of

$$R = R_0\left(1+a\frac{t}{T}\right) \tag{8-23}$$

for different parameters a.[2,3] It is seen that for $a\geq2$, the mean value of the output voltage at the time instant $t=T$ approximates the final value with a maximum error of 3.7 instead of 13.5 percent.

The continuously varying resistance may be realized by one of the following methods:

1. Motor driven potentiometer[20]
2. Indirectly heated thermistor[3]
3. Photoresistance with varying illumination[57]

The stepwise varying resistance may be most simply realized using an electronically switched additional resistance network[4,5,12,58] (see Fig. 8-6).

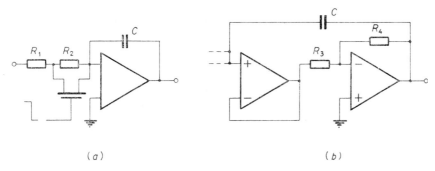

(*a*)   (*b*)

Figure 8-6 *(a)* Integrating circuit with varying time constant switched by a MOS transistor. *(b)* Increasing the time constant by a Miller capacitor $\tau=(R_1+R_2)C(1+R_4/R_3)$.

It can be shown using computer simulation that a single step of resistance change near the time instant $t=T/2$ is sufficient to meet average accuracy requirements in the circuit shown in Fig. 8-6.[4,5] In Fig. 8-7, output signals of networks with constant and varying time constants are compared for linearly detected normal distribution input signals.

The reduction in the price of analog computing networks (operational amplifier, multipliers, and dividers) has made feasible the realization of the relation

$$M(\eta)_T = \frac{1}{T} \int_0^T \eta(t)\, dt \qquad (8\text{-}24)$$

similar to Eq. (8-20), for varying, but not infinitely long, time intervals T.[60] This has the advantage that $M(\eta)$ may be read off at any time; its accuracy increases with the elapsed time. According to Fig. 8-8*a* the output signal, which is proportional to the time integral of the input signal, is divided by a quantity corresponding to the time interval T that has elapsed from the beginning of the integration using an analog divider circuit. T may not be increased arbitrarily because of the danger of overflow. However, this may also be eliminated by the arrangement shown in Fig. 8-8*b* according to which the divider and integrator are driven by the difference between the input and output signals. For $T \leq T_0$, the divider input is driven by T_0 in order to eliminate division by zero. Figure 8-8*c* shows the time dependence of the output signal.

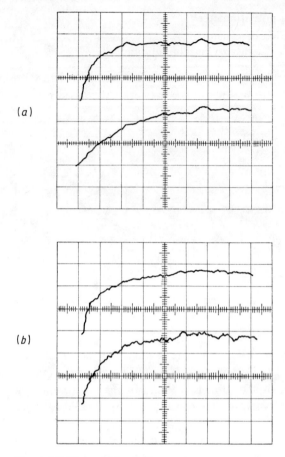

Figure 8-7 Noise driven *RC* network output responses.
(a) Varying parameter network (upper trace), network
which is equivalent when reaching the stationary state
(lower trace). *(b)* Varying parameter network (upper
trace); network which is equivalent at the starting time
instant (lower trace).

Similar circuit elements may be utilized to realize a square-law detector comprising an integrator,[46] which has the advantage of showing the root mean square
value on a linear scale. According to definition

$$D_T^2(\xi) = \frac{1}{T} \int_0^T \xi^2(t)\, dt \tag{8-25}$$

where the subscript T denotes the dependence on integration time. Calculating
the derivative

$$\frac{d}{dT}[TD_T^2(\xi)] = 2TD_T(\xi)\frac{dD_T(\xi)}{dT} + D_T^2(\xi) = \xi^2(T) \tag{8-26}$$

we have, after rearrangement,

$$\frac{dD_T(\xi)}{dT} = \frac{1}{2T}\left[\frac{\xi^2(T)}{D_T(\xi)} - D_T(\xi)\right]$$ (8-27)

The analog circuit model of this expression is shown in Fig. 8-9. Integrators with varying time constants have recently come into widespread use.[56]

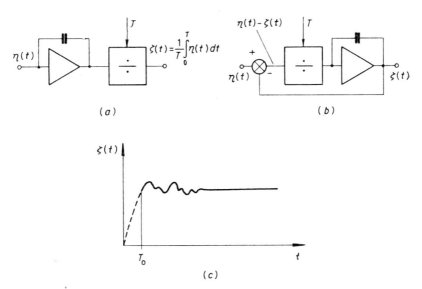

Figure 8-8 (*a*) An ideal integrator with varying time constant.
(*b*) Arrangement that is free from overflow. (*c*) Time function of output signal.

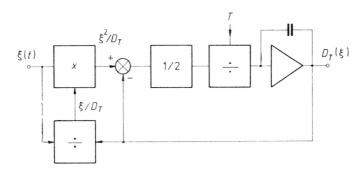

Figure 8-9 Integrator with square-law operation but providing linear indication.

8-3 DIGITAL INTEGRATION

Due to drift effects and the imperfections of analog storage devices, the analog principle is not suitable for meeting high-accuracy requirements or for performing narrow-band and low-frequency measurements. Evidently, the components of analog systems presented in Sec. 8-2 may be substituted by digital components. However, there are measurement systems inherently realized by digital circuits (e.g., for counting the crossing of threshold levels) not requiring previous detection.

Figure 8-10a shows the possible time function of a white-noise signal with normal distribution and band-limited to a narrow frequency range. For a band-center frequency f_0, the number of zero crossings in the increasing direction will be f_0. The frequency of crossing the level U_0 will be $e^{-U_0^2/2\sigma_u^2}$ times f_0 which can be seen at once from knowledge of the normal distribution density function. During time interval T, the expected number of all zero crossings (e.g., in the positive direction)[68] is

$$M(N) = Tf_0 e^{-U_0^2/2\sigma_u^2} \tag{8-28}$$

and their variance is

$$D^2(N) = \frac{e^{U_0^2/2\sigma_u^2} - 1}{2\Delta fT} \quad \text{if} \quad U_0 \cong \sigma_u \tag{8-29}$$

Knowledge of U_0 and N allows the calculation of σ_u. The value of f_0 in Eq. (8-28) is preferably determined by counting the zero crossings, as f_0 depends also on the shape of the band-pass filter response [see Eq. (5-98)].

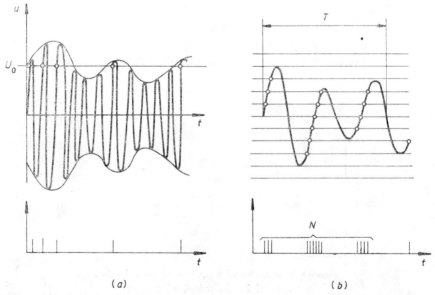

(a)   (b)

Figure 8-10 (a) Limit exceeding time instants of band-limited white noise. (b) Multilevel counting. (The number of periods during time interval T is given by $n = Tf_0$.)

This dependence, in the general form,[55] is

$$f_0 = \sqrt{-\frac{\Gamma''(0)}{\Gamma(0)}} = \left[\frac{\int\limits_0^\infty f^2 S(f)\,df}{\int\limits_0^\infty S(f)\,df} \right]^{\frac{1}{2}} \qquad (8\text{-}30)$$

where $\Gamma''(\tau)$ is the second derivative of the autocorrelation function pertaining to the band-limited signal. An easily applicable formula for calculating the variance of the number of zero crossings has only recently been derived.[15]

The counting of single level crossings may only be applied for $U_0 \approx \sigma_u$. If σ_u changes within a wide range, multilevel counting is preferred[62] (see Fig. 8-10b). It may be seen from Fig. 8-10b that for N level crossings in the positive direction during n periods, the average peak-to-peak voltage w il be

$$M(U_{p-p}) = \frac{N}{n}\,\Delta U \qquad (8\text{-}31)$$

which is just twice the expected value of the Rayleigh distribution envelope. Utilizing Eq. (2-141)

$$\sigma_u = \sqrt{\frac{2}{\pi}\frac{M(U_{p-p})}{2}} = \frac{\Delta U}{\sqrt{2\pi}}\frac{N}{n} = \frac{\Delta U}{\sqrt{2\pi}}\frac{N}{Tf_0} \qquad (8\text{-}32)$$

For a wider frequency range or a complicated $S(f)$ function, the determination of f_0 is difficult.[19] However, the counting of level crossings may also be applied to a stochastic signal band-limited in one direction only, i.e. after transmitting it through a low-pass network. This is possible by the application of an inserted integrator,[63] or when $S(f)$ in Eq. (8-30) is a polynomial of f and above a certain frequency f_1, $S(f)$ decreases more rapidly than $1/f^2$.[1]

By digitally processing an amplitude quantization of $2n$ levels, the mean value, the mean square value, and the amplitude density function of a signal having arbitrary time function may be determined with arbitrary accuracy.[39] According to Fig. 8-11, the absolute value of the signal is less than $|U/2n|$ in the interval $T-t_1$, less than $|3U/2n|$ in the interval t_1-t_2, etc. The values of T and t_i, measured digitally with high accuracy, may be substituted into the relation

$$\int\limits_0^T u^2\,dt \cong (T-t_1)\left(\frac{U}{2n}\right)^2 + (t_1-t_2)\left(\frac{3U}{2n}\right)^2 + \ldots + (t_{n-1}-t_n)\left(\frac{2n-1}{2n}U\right)^2$$

$$= T\frac{U^2}{4n^2} + \frac{2U^2}{n^2}\sum_{i=1}^{n-1} it_i \qquad (8\text{-}33)$$

For $n=16$, the error of the root mean square value is less than 0.5 percent. An even more simple relationship, similar to Eq. (8-32), can also be derived for the expected value.

The sampling procedure illustrated in Fig. 8-12a may generally be applied

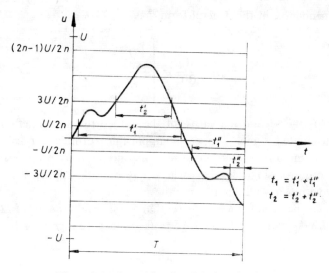

Figure 8-11 Quantizing for digital evaluation.

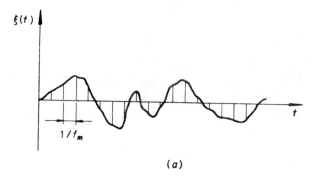

(a)

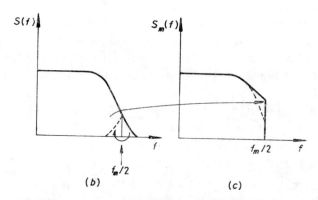

(b)   (c)

Figure 8-12 (a) Sampling of stochastic signal. (b) Original power spectrum. (c) Power spectrum distorted by the loss of signal components above $f_m/2$.

with due caution. If components of the investigated signal which are higher than $f_m/2$ are passed through the filter (see Fig. 8-12b), then the spectrum determined after sampling will not be true (see Fig. 8-12c). This effect may be counteracted by effective band-limiting (i.e. using a filter with sharp cut-off) or by increasing f_m.

The samples of the analog signal shown in Fig. 8-12 have to be quantized before further digital processing. This results in an inevitable error which may be decreased by increasing the resolution of the analog/digital (A/D) converter. Another limitation is introduced by the finite conversion speed of the converter: f_m may not be increased arbitrarily.

Several methods of further processing the digital signal samples are known.[45,69] In addition to simple algorithms, which may be implemented using any computer, there are special procedures such as that suitable for determining the autocorrelation function.[51] From this, the power spectrum may be calculated using the Fourier transform method. This method has become feasible with the advent of programs for performing the fast Fourier transform (FFT).[49]

Evidently, the FFT is also suitable for determining the direct Fourier transform of stochastic signal samples. The function thus determined is, of course, not continuous but is made up of a series of discrete samples. The convergence problems mentioned in Chapter 3 [see Eq. (3-22)] have no significance here as only processes of finite length may be analyzed so the requirement given in Eq. (3-23) is automatically met. It follows from the algorithm of the FFT[72] that the frequency axis of the spectrum calculated will have a linear scale.

In many cases, the logarithmic frequency scale of constant relative frequency increments, which requires digital integration, is more suitable. The filters shown in Fig. 8-1 are then preferably realized as digital ones:[37,67] these comprise time delay networks. The delay time may be adjusted by altering the clock frequency, without changing the filter characteristic on a logarithmic scale. During this process, the band-center and cut-off frequencies are directly proportional to the clock frequency.

Considering digital measurement methods, the method that employs stochastic–ergodic pulse techniques has only small hardware requirements.[70] The input analog signal is converted into a stochastic series of logic 0 and 1 pulses by a stochastic converter. The basic operations (i.e. addition, subtraction, and multiplication) needed for the evaluation are performed by simple logic gates, the time shift is implemented by a shift register, and integration is achieved using a counter.

8-4 MEASUREMENT OF DENSITY FUNKTIONS

The measurement of the amlitude density function, the power spectrum, and the autocorrelation function presents similar problems. According to Fig. 8-13, windows of width Δx, Δf, and $\Delta \tau$ have to scan continuously, or stepwise, the range of independent variables x, f, and τ, and the function to be determined may be expressed as the series of measured values. In order to increase the resolution, the window width should be decreased. However, this would result either in increased measurement time or—because of the stochastic nature of the measured data—a decreased accuracy. The original function may also be reconstructed from a series of data measured with a wide window,[66] but the accuracy of this procedure is limited by stochastic disturbances of the measuring equipment, even during the reconstruction of a deterministic function.

Figure 8-14 shows the simplified block diagram for the measurement of the amplitude density function. From the point of view of the circuit, the shift in the

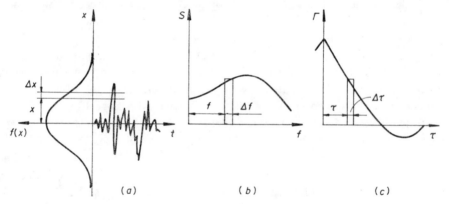

(a) (b) (c)

Figure 8-13 Principle of measurement: *(a)* Amplitude density function. *(b)* Power spectrum. *(c)* Autocorrelation function.

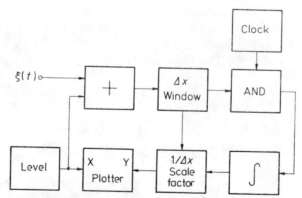

Figure 8-14 Simplified block diagram for the measurement of the amplitude density function.

window level is not generally feasible, so the mean value of the signal under test is shifted by a superimposed variable DC voltage. The window circuit, made up of operational amplifiers, comparators, or Schmitt triggers,[10, 17, 24, 28, 50, 71] is essentially a two-state flip-flop. The window frequently is realized by a narrow light-transmitting slot and a photomultiplier positioned in front of a cathode ray tube screen, or an electrode fastened to the screen.[18, 31, 42] If the instantaneous value of the signal under test falls between the two levels, the AND gate passes the pulses of the clock generator. One input of an XY recorder is driven by the average value of the pulses, while the other input senses the DC voltage superimposed on the signal under test.

According to Eq. (2-14), the probability of the instantaneous signal value falling into the window of width Δx is

$$\frac{\Delta t}{T} = f(x)\, \Delta x \tag{8-34}$$

Here T is the time needed to measure $f(x)$ pertaining to a fixed x, Δt is the time spent within the window, and $f(x)$ is the density function to be determined. Again, a stochastic signal of bandwidth Δf and of normal distribution is assumed. The number of independent samples arriving during time interval Δt is $2\Delta f T f(x)\Delta x$ (see Sec. 8-1). Evidently, this value fluctuates and its distribution is generally complicated or not known.[19] The relative standard deviation

$$\psi_a = \frac{D}{M}[2\Delta f T f(x)\, \Delta x] = \frac{A}{\sqrt{\Delta f T f(x)\, \Delta x}} \tag{8-35}$$

introduced as an analog quantity to Eqs. (8-6) and (8-7), gives information on the factors influencing the accuracy of the measurement. In the numerator, A is a factor depending on $f(x)$, x, and Δx, its value falling between 0.15 and 0.7.[14, 55] It can be seen from Eqs. (8-34) and (8-35) that for small $f(x)$ values (i.e. at the extremes of the gaussian distribution) the number of samples that may be regarded as independent is also small and the relative fluctuation is high. This may be altered by widening Δx (by decreasing the resolution) or by increasing the measurement time.

The above considerations are valid for the determination of $f(x)$ pertaining to a single x value. If the complete density function has to be scanned, the measurement time is further increased because the scanning speed may not be increased above $\Delta x/T$ without impairing the accuracy. The complete measurement time is thus

$$T_t \gtrsim \frac{x_{max} - x_{min}}{\Delta x} T \tag{8-36}$$

where x_{max} and x_{min} are the extreme abscissa values of the density function which have to be investigated. In case of several windows operating in parallel, the complete measurement time is governed by the measurement time pertaining to the worst case.[38]

The simplified block diagram of a spectrum analyzer suitable for showing the power spectrum is presented in Fig. 8-15. The window is realized by a narrow band-pass filter; and the signal under test, by mixing it with a local oscillator signal with a continuously or stepwise variable frequency, is shifted in front of a window of fixed intermediate frequency.[22] According to another method applied primarily to the audio frequency range, a series of windows tuned to different frequencies is used. The signal under test is applied to the parallel-connected input terminals of these windows, and the power spectrum is generated by scanning the output terminals.[23, 59] Any of these methods is suitable for measuring the one-sided spectrum $S(f)$ or $S(\omega)$ in a range of positive frequencies; according to Chapter 3, $s(f)$ or $s(\omega)$ is determined by halving.

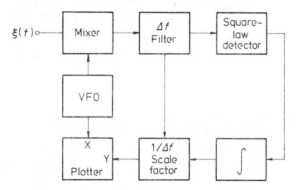

Figure 8-15 Simplified block diagram of power spectrum analyzer. In the case of varying bandwidth Δf, the scale factor has to be varied simultaneously. VFO: variable frequency oscillator.

The first method is characterized by a constant absolute bandwidth, while the second has a constant relative bandwidth. Both are realizable using either analog or digital means (see Sec. 8-3). The attainable speed of highly complex digital integrated circuits at present is not sufficient for application to digital spectrum analysis at arbitrarily high frequencies. Analog methods of spectrum analysis, although being replaced at an increasing rate by digital methods, will probably remain in widespread use at higher frequencies for a long time to come.

Applying a signal of relative bandwidth Δf to a square-law detector and performing ideal integration during a time interval T, the result of integration will show the relative standard deviation given by Eq. (8-6). Applying RC filtering following the detection, the relative fluctuation will be given by Eq. (8-18). In this case too, the accuracy attainable is inversely proportional to the product of the window width Δf and the measurement time T.

The power spectrum of a signal having a known amplitude density spectrum may also be calculated from the measurement performed using a linear detector. According to Eqs. (8-7) and (8-19), respectively, the relative error of the output signal supplied by a linear detector with an ideal integrator or an RC filter is approximately half the error given by a square-law detector. However, this does not give a more accurate result because the relative error is doubled by squaring the data obtained by linear detection.[64]

For the analysis of a wide frequency range, the required time, as in Eq. (8-36), is

$$T_t \cong \frac{f_{max} - f_{min}}{\Delta f} T \tag{8-37}$$

In the case of constant relative resolution $(\Delta f/f = \text{const.}$, which is easier to realize with nonheterodyne-type analyzers), Δf is increased and T_t is decreased by increasing frequency. Thus

$$T_t \cong (f_{max} - f_{min}) T \sum_{1}^{n} \frac{1}{\Delta f_i} < \frac{f_{max} - f_{min}}{\Delta f} T \tag{8-38}$$

Further improvement is achieved by varying the time T_i needed for the measurement of differing Δf_i frequency bands, according to the law $\Delta f_i T_i = \text{const.}$

Evaluation of spectrum analysis results requires exact knowledge of the band-center and bandwidth data of the band-pass filters. However, according to Sec. 5-5, these are also dependent on the spectrum to be measured, so the prewhitening of the spectrum using suitable emphasis networks is sometimes applied.[16, 45] The detailed analysis of flicker noise requires an equalizer which has a transfer function given by $|H(\omega)| = \text{const.} \sqrt{\omega}$.[33, 43]

Finally, the method for measuring the autocorrelation function is shown in Fig. 8-16. The time function $\xi(t)$ and its delayed version $\xi(t-\tau)$ are multiplied, and the average value of the product is used to drive the Y input of an XY recorder[53]. The X deflection is proportional with τ.

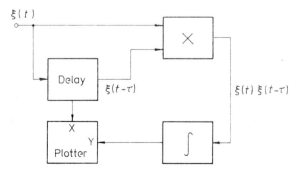

Figure 8-16 Simplified block diagram of autocorrelator.

If $\xi(t)$ has normal distribution and is limited to the frequency range Δf, the relative standard deviation of the autocorrelation function[14] is

$$\psi_c = \frac{D[\Gamma_\xi(\tau)]}{M[\Gamma_\xi(\tau)]} = \frac{1}{\sqrt{2\Delta f T}}\left[1+\frac{\Gamma_\xi^2(0)}{\Gamma_\xi^2(\tau)}\right] \tag{8-39}$$

$\Gamma_\xi(\tau)$ is decreased by increasing τ. This means that, as in the relation (8-35) which applies to the relative standard deviation of the amplitude density spectrum, the number of samples per unit time that may be considered independent is also decreased, so ψ_c is increased.

Resolution is governed by the sampling period $\Delta\tau$. As $\Gamma_\xi(\tau)$ and $s(f)$ are Fourier transforms of each other, the upper limit of the equivalent power spectrum is limited to $f_{max}=1/2\Delta\tau$ by the choice of $\Delta\tau$. There is also a lower limit. It follows from the definition of the autocorrelation function

$$\Gamma_\xi(\tau) = \frac{1}{T}\int_0^T \xi(t)\xi(t-\tau)\,dt \tag{8-40}$$

that sufficient accuracy in the measurement of $\Gamma_\xi(\tau)$ is only achievable in the range $0<\tau\ll T$. So the components with lower frequencies than $1/2\tau$ are lost.

The power spectrum measured according to Fig. 8-15 and the autocorrelation function measured according to Fig. 8-16 can only be perfectly equivalent if $f_{max}-f_{min}$ and τ tend to infinity. The functions practically measured are always truncated, so the Fourier transforms applying to the actual function are only approximated, by a finite error, using the measured transforms.[16] The two methods may be considered to be supplementary, and neither can be used for substituting the other. Thus at high frequencies, the power spectrum is readily measured, while at frequencies below 1 Hz the measurement of the autocorrelation function is more feasible.

8-5 MEASUREMENT OF ONE-PORT AND TWO-PORT NOISE PARAMETERS

The linear and square-law average value measuring devices, amplitude and spectrum analyzers, etc. considered in the previous Sections are theoretically suitable for the measurement of any noise parameter of any network, if supplemented by a preamplifier of sufficient sensitivity and a band-pass filter. However, low-level measurements present several problems, especially when considering one-ports and low-gain two-ports.

The noise voltage of a forward biased semiconductor diode would be difficult to measure accurately according to the arrangement shown in Fig. 8-1, or any other similar simple arrangement, as the preamplifier noise voltage, reduced to the input, may be higher than the noise voltage to be measured. Two methods can be applied to overcome this difficulty: the use of an extremely low-noise pre-amplifier or the elimination of the effect caused by the additional noise voltage by a substitution method.

As shown in Problem 6-1, the input stage has the lowest noise figure at low and medium frequencies. This is achieved if a p–n junction field-effect transistor is applied, especially for signal sources of high internal resistance (10–100 kΩ).

The two-channel amplifier with correlator[21] shown in Fig. 8-17 has extremely low noise. The noise to be measured is amplified by both channels, and due to perfect correlation, the mean square value is directly obtained at the correlator output. If the equivalent noise generators of the two input stages are independent of each other, their cross correlation is zero and they have no effect at the output. Unfortunately, the same is not true for current generators; in spite of their independence they are mutually coupled through the device under test. This means that devices having extremely low input noise currents are needed for the correlating amplifier shown in Fig. 8-17.

In a frequency range of less than a few hundred Hz, the flicker noise of the measuring amplifier presents additional problems. According to the arrangement utilizing a modulator shown in Fig. 8-18, the flicker noise is eliminated by the pass-band range of the amplifier which has to cover only the switching frequency of several kHz and a few harmonics. The device, a BJT or a field-effect transistor,

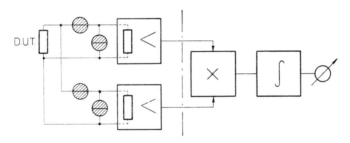

Figure 8-17 Simplified block diagram of a low-noise amplifier utilizing a **correlator**.

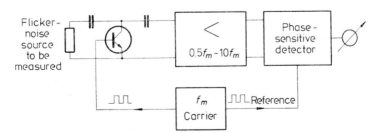

Figure 8-18 Simplified block diagram of low-frequency noise measurement utilizing a modulator.

bridging the input terminals, is biased either to cut-off or into the saturation region. In both cases, its flicker noise is negligible.[13]

A perfect elimination of the preamplifier noise may not be achieved by any of the methods shown, but the error thus introduced may be decreased by choosing a suitable measuring principle. The equivalent noise current of a semiconductor diode may be accurately measured using the arrangement shown in Fig. 8-19. First, the condition $Y_d = G + jB$ has to be adjusted at the measuring frequency, by a bridge circuit not shown. Next, the saturation currents of vacuum diodes D_1 and D_2 are changed to obtain the same output signal at the indicator amplifier output in the two positions of switch S. Due to the comparison method used, the result is not affected by the amplifier noise.[26]

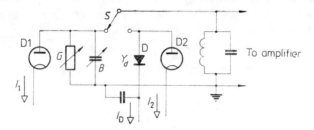

Figure 8-19 Substitution method for one-port noise measurement. D is the diode under test.

Special attention should be given to the power supply feeding the device under test or the preamplifier. Batteries or storage batteries may be used to eliminate disturbances originating from the mains supply. The flicker noise of fresh batteries is even negligible.[40]

The noise parameters of an active two-port are generally more easily measurable as the power gain A_p of the two-port has the effect of decreasing the noise of the measuring amplifier by a factor $1/A_p$. It has been shown in Sec. 5-2 that from any two-port, two-noise generators, which are generally partially correlated, may be extracted. These are characterized by four parameters, so at least four measurements are needed for their determination (see Fig. 8-20). The measurement of u^l and i^l is self-explanatory: the correlation factor characterizing their relation has a real part a and an imaginary part b which, according to Eq. (5-46), may be calculated from the relation

$$|u_i|^2 = \left|\frac{Z_i}{Z_i + Z_g}\right|^2 \left[|u_g|^2 + |u^l|^2 + |Z_g i^l|^2 + 2\sqrt{|u^l|^2 |i^l|^2}(aR_g - bX_g)\right] \quad (8\text{-}41)$$

in case of termination, $X_g = 0$ and $R_g \neq 0$ or $R_g = 0$ and $X_g \neq 0$, respectively. Theoretically, R_g and X_g may be chosen arbitrarily; however, this choice affects the accuracy of the measurement.

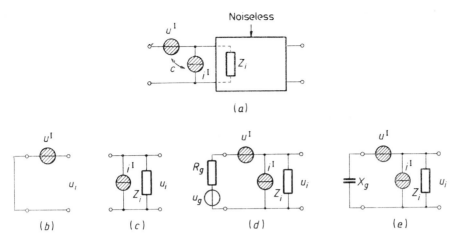

Figure 8-20 *(a)* Two-port with noise sources reduced to the input. Measurement principles of: *(b)* u^I source voltage. *(c)* i^I source current. *(d)* Real component a of the correlation factor. *(e)* Imaginary component b of the correlation factor. [See Eq. (8-41).]

The measurement of the four noise parameters mentioned above requires knowledge of the two-port gain. This is not feasible for every case, so the feedback measurement systems shown in Fig. 5-15 are generally applied. If the overall amplification of the device under test and the measuring amplifier are sufficiently high to introduce a loop gain significantly greater than one, this overall amplification is determined by the feedback network only. Naturally, in accordance with Sec. 5-4, the feedback network should not comprise substantial noise sources[11].

In spite of the fact that u^I, i^I, a, and b are necessary and sufficient parameters for noise characterization, actually a fifth parameter—the noise figure is generally measured. The reason for this is the simplicity of the technique and the meaningful interpretation of the noise figure.

8-6 MEASUREMENT OF THE NOISE FIGURE[9]

According to Eq. (5-39), the noise figure may only be defined if the signal source feeding the two-port input has an impedance comprising a resistive component. In order to obtain definite measurement results, the signal source is generally purely resistive; otherwise matching conditions should be clearly defined.[44, 47, 52] This is explained by the fact that according to Eq. (5-61), the noise figure has a minimum as a function of both components of the signal source admittance $Y_g = G_g + jB_g$. The signal source admittance yielding the minimum noise figure is not necessarily equal to that providing optimum power match.

From the parameters measured according to the general arrangement shown in Fig. 8-21, the noise figure may be calculated from the relationship

$$F = \frac{u_0^2}{A_p 4kT R_g \Delta f} \tag{8-42}$$

$4kT$, R_g, and Δf are quantities previously measured or adjusted, but u_0^2 and A_p depend on the two-port under test. These may be measured using several methods.

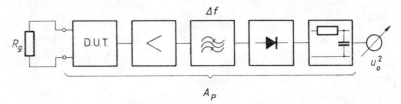

Figure 8-21 Simplified block diagram of noise figure measurement.

Separate Measurement of Noise Power and Amplification

In the arrangement shown in Fig. 8-22a, amplification is determined by applying to the input a sine-wave signal of relatively high amplitude, and inserting a known attenuation into the amplifier chain to avoid detector overload. According to

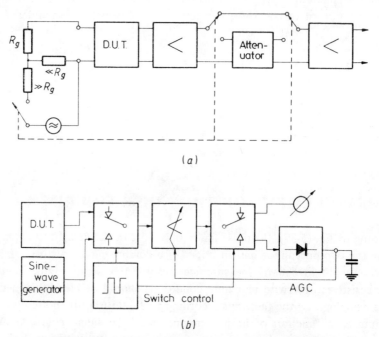

(a)

(b)

Figure 8-22 (a) Gain measurement with high-amplitude sine-wave signal. (b) Automatic gain control (AGC).

the arrangement shown in Fig. 8-22b, the amplification is kept constant automatically by inserting the measuring and control channels alternately using an
electronic switch (time division multiplex-type of gain control). On the other
hand, frequency division multiplex-type of gain control can be achieved by
introducing a pilot signal of a frequency that is outside, but near to, the measurement frequency range Δf.[38]

Noise Figure Measurement with Known Amplification

In some cases, the gain of the two-port under test is precisely known (e.g., transistor in common base configuration[8]), then the remaining amplifier stages may
be stabilized by negative feedback. On these occasions, it is sufficient to measure u_0^2.

Increase in Output Power Obtained with a Sine-Wave Input Signal of Known Amplitude

The arrangement shown in Fig. 8-22a may be utilized, but switches and an attenuator are not needed. The root mean square value of the sine-wave input signal
is increased until the output power is increased by a known amount (e.g., doubled).
Thus

$$(u_0')^2 = A_P 4kTR_g \Delta f + A_P u_{\sin}^2 = 2(u_0)^2 \tag{8-43}$$

Comparing this with Eq. (8-42), we have

$$F = \frac{u_{\sin}^2}{4kTR_g \Delta f} \tag{8-44}$$

This method may only be applied if the detector has a square-law characteristic
or if its response to input signals having different amplitude density functions
is known.

Increase in Output Power Obtained with an Input Noise of Known Power

According to the arrangement shown in Fig. 8-23a, the input terminating resistance
R_g is driven by a source supplying a calibrated noise current (e.g., a vacuum diode
operating in the saturation region — see Sec. 4-1). The output power, in the case
of doubling, is

$$(u_0')^2 = A_P 4kTR_g \Delta f + A_P 2qI\Delta fR_g^2 = 2(u_0)^2 \tag{8-45}$$

which yields

$$F = \frac{q}{2kT} IR_g$$

(8-46)

As anticipated, the bandwidth Δf does not appear in this formula, so a filter with arbitrary transfer function may be applied. If the noise to be measured and the calibrated noise both have normal distribution, u_0 may be unambiguously measured even using a linear detector.

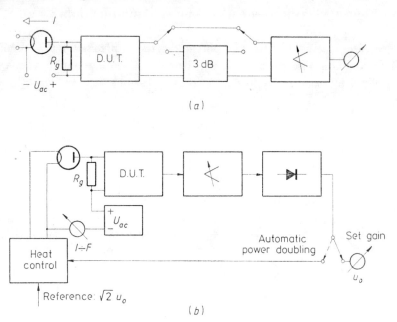

Figure 8-23 *(a)* Noise figure measuring set with calibrated noise source and inserted attenuator. *(b)* Semiautomatic noise figure measuring set.

Theoretically, no flicker noise is generated by a saturated vacuum diode having a pure metallic cathode. However, a cathode is not perfectly pure due to contaminants within the tube envelope and also due to the imperfect vacuum. As a result flicker noise may be introduced. This is why at low frequencies, especially since the widespread introduction of digital techniques, artificial noise sources supplying a pseudo-random binary signal are generally applied.[25, 29, 48, 65]

The measurement method based on output power multiplication has several variants. According to Fig. 8-23a, a 3-dB attenuator may be inserted between the two-port under test and the amplifier. This attenuator halves the power increment supplied by the additional noise source, so that during the measurement, the same deflection has to be adjusted on the meter showing u_0^2 under both conditions. It can be shown that if the 3-dB attenuator is connected directly to the output of the device under test then the measuring amplifier noise does not introduce any measurement error.[54]

Of the two measurements needed for the noise figure determination, one or both may be assessed automatically. In Fig. 8-23b, the output voltage is first adjusted to a prescribed value by adjusting the gain control, and the power doubling is set by an automatic control circuit.[6,41] The noise figure, according to Eq. (8-46), is proportional to the saturation current I of the diode.

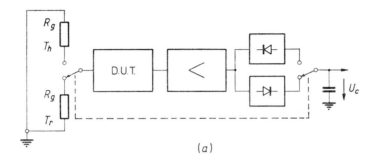

(a)

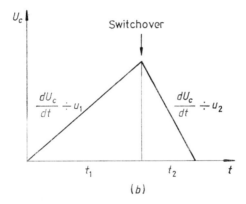

(b)

Figure 8-24 *(a)* Simplified block diagram of high-accuracy noise measuring set operating with power multiplication. *(b)* Time function of integrator output voltage. The second zero crossing is sensed by a comparator not shown.

If the adjustment of the input noise generator power presents difficulties, the input noise is kept constant, and the factor by which the output noise is increased due to this constant input noise is measured. This principle is applied to the high-accuracy transistor noise measuring apparatus operating in the audio frequency range (see Fig. 8-24).[7] The calibrated noise source is a heated resistance. First, the input of the transistor under test is connected to the signal source at room temperature. For a noise figure F, the noise voltage reduced to the input is

$$u_1^2 = F u_r^2 \tag{8-47}$$

where u_r is the noise voltage given by the cold resistance (at room temperature). Next, the transistor is switched over to the heated resistance when

$$u_2^2 = F u_r^2 + \Delta u_h^2 \tag{8-48}$$

Here

$$\Delta u_h^2 = u_h^2 - u_r^2 = u_r^2 \left(\frac{T_h}{T_r} - 1 \right) \tag{8-49}$$

is the additional squared noise voltage of the heated resistance. T_h and T_r are the absolute temperatures of the resistances. Mutual substitution of the above relations for $T_h = 3T_r$ yields

$$\frac{u_2}{u_1} = \sqrt{\frac{F+2}{F}} \tag{8-50}$$

and

$$F = 10 \log \frac{2(u_1/u_2)^2}{1 - (u_1/u_2)^2} \quad \text{dB} \tag{8-51}$$

respectively. According to Fig. 8-24, the currents proportional to u_1 and u_2 are utilized to charge and discharge a capacitor, respectively. The digital measurement of the charge and discharge time allows the determination of the noise figure with an accuracy of ± 0.2 dB.

Several noise sources generating white noise, flicker noise, and burst noise simultaneously may be separated by more sophisticated methods.[27,61]

PROBLEMS

8-1 A power spectrum has to be measured in the frequency range 10 Hz to 10 kHz with a constant bandwidth of 10 Hz and a relative standard deviation of 10 percent. What time is needed for the spectrum measurement?

Answer: Using an ideal integrator, the measurement of a single band of 10 Hz width requires, according to Eq. (8-6), a measurement time of

$$T = \frac{1}{\psi_s^2 \Delta f} = \frac{1}{10^{-2} \times 10} = 10 \text{ s}$$

Thus the measurement of the complete frequency range requires a measurement time of

$$T_t = \frac{f_{max} - f_{min}}{\Delta f} T = \frac{9990}{10} 10 = 9990 \text{ s}$$

8-2 What is the time requirement when the complete frequency range is scanned with a relative bandwidth of $\Delta f/f = 0.2$?

Answer: In order to cover the range 10 Hz to 10 kHz, the solution of the equation

$$1.2^n = \frac{10^4}{10}$$

yields $n = 38$ filter windows. (In the previous case, this number was 999.) The measurement of a single window requires

$$T = \frac{1}{\psi_s^2 \Delta f} = \frac{5}{\psi_s^2 f}$$

so the measurement of the complete frequency range requires

$$T_t = \sum_{i=1}^{n} T_i = \frac{5}{\psi_s^2} \sum_{i=1}^{n} \frac{1}{f_i} = \frac{5}{\psi_s^2 f_{min}} \sum_{i=1}^{n} \left(\frac{1}{1.2}\right)^n < \frac{5}{\psi_s^2 f_{min}} \frac{1}{1 - 1/1.2} = 300 \text{ s}$$

This great advantage may not be utilized in practice because the switching of the integrator (or low-pass filter) for each band presents difficulties.

REFERENCES

[1] Aitken, G. J. M. and J. M. Cruickshank: "The Direct Digital Measurement of Low-Pass Noise Power", *IEEE Trans.*, vol. IM-21, no. 3, pp. 255—258, 1972.

[2] Ambrózy, A.: "Reducing the Time Requirement in Direct-Reading VLF Noise Measurement", *Proc. IEEE*, vol. 53, no. 8, pp. 1161—1162, 1965.

[3] —: "Reducing the Time Requirement in Direct-Reading Noise Measurement", *Period. Polytechn.*, vol. 9, no. 3, pp. 301—320, 1965.

[4] —: "Time-Varying Circuits for Noise Measurement", *Proc. IEEE*, vol. 56, no. 1, pp. 78—79, 1968.

[5] —: "Some New Time-Varying Smoothing Circuits for Direct-Reading Noise Measurement", *Period. Polytechn.*, vol. 12, no. 1, pp. 61—70, 1968.

[6] —: "Semiautomatic High-Frequency Transistor Noise Measurement Set", (Félautomatikus nagyfrekvenciás tranzisztor zajmérő.) *Mérés és Automatika*, vol. 16, no. 12, pp. 492—493, 1968, (in Hungarian).

[7] —: "Accurate Low-Frequency Transistor Noise-Measuring Instrument with Digital Display of Logarithmic Noise Figure", *Period. Polytechn.*, vol. 15, no. 4, pp. 371—379, 1971.

[8] — *et al.*: "Direct-Reading Transistor Noise Figure Measuring Instrument", *Electr. Eng.*, vol. 35, no. 9, pp. 611—613, 1963.

[9] Arthur, M. G.: "The Measurement of Noise Performance Factors: A Metrology Guide", *National Bureau of Standards Monograph*, no. 142, Washington, 1974.

[10] Aumüller, W. and A. Schief: "Ein Amplitudenanalysator zur Ermittlung statistischer Eigenschaften kontinuierlich verlaufender Signale", *Elektr. Rdsch.*, vol. 16, no. 6, pp. 249—253, 1962.

[11] Ayre, R. W. A.: "A New Transistor Noise Test Set", *Proc. IEEE*, vol. 60, no. 1, p. 151, 1972.

[12] Ball, E.: "Realization of the Mean-Value Function", *Electr. Lett.,* vol. 9, no. 22, pp. 528—529, 1973.

[13] Baxandall, P. J.: "Low-Noise LF Amplification Using a Transistor Chopper", *Electr. Lett.,* vol. 4, no. 1, pp. 14—15, 1968.

[14] Bendat, J. S. and A. G. Piersol: *Measurement and Analysis of Random Data,* John Wiley, New York, 1966.

[15] Blachman, N. M.: "Narrow-Band Gaussian Noise: Variance of the Number of Zero Crossings", *Electr. Lett.,* vol. 14, no. 3, pp. 75—76, 1978.

[16] Blackman, R. B. and J. W. Tukey: "The Measurement of Power Spectra", *BSTJ,* vol. 37, no. 1, pp. 185—282; no. 2, pp. 485—569, 1958.

[17] Blackwell, C. A. and R. S. Simpson: "Analog Method for Determining Probability Distribution of Random Signals", *Rev. Sci. Instr.,* vol. 36, no. 12, pp. 1877—1878, 1965.

[18] Broch, J. T.: "Automatic Recording of Amplitude Density Curves", *Brüel–Kjaer Tech. Rev.,* vol. 6, no. 4, pp. 3—19, 1959.

[19] —: "Peak Distribution Effects", *Brüel–Kjaer Tech. Rev.,* no. 1, pp. 3—31, 1968.

[20] Brooks, H. B. and K. E. Walker: "Average Responding Instruments", *IRE Trans.,* vol. IM-6, no. 4, pp. 258—260, 1957.

[21] Brophy, J. *et al.*: "Correlator Amplifier for Very Low Level Signals", *Rev. Sci. Instr.,* vol. 36, no. 12, pp. 1803—1806, 1965.

[22] Brüel and Kjaer Data Sheet: "Frequency Analyzer 2010".

[23] Brüel and Kjaer Data Sheets: "Audio Frequency Spectrometers 2120, 2131".

[24] Caldwell, W. F. *et al.*: "A Precision Amplitude Distribution Amplifier", *IRE Trans.,* vol. EC-9, no. 6, p. 252, 1960.

[25] Castanie, F.: "A Uniformly Distributed Analog Random Voltage Generator", *Proc. IEEE,* vol. 66, no. 5, pp. 605—606, 1978.

[26] Champlin, K. S.: "Bridge Method of Measuring Noise in Low-Noise Devices at Radio Frequencies", *Proc. IRE,* vol. 46, no. 4, p. 779, 1958.

[27] Choe, H. M. and A. van der Ziel: "Discriminating between Noise Sources by Modulation Techniques", *Solid State Electr.,* vol. 19, no. 8, p. 737, 1976.

[28] Clarke, K. K.: "An Electronic Probability Density Machine", *IEEE Trans.,* vol. IM-15, nos 1—2, pp. 25—29, 1966.

[29] Davies, A. C.: "Probability Distributions of Noiselike Waveforms Generated by a Digital Technique", *Electr. Lett.,* vol. 4, no. 19, pp. 421—423, 1968.

[30] Dietzel, R.: "Anzeigeschwankungen bei der Effektivwertmessung von Bandpassrauschen", *Hochfrequ. u. El. ak.,* vol. 72, no. 10, pp. 172—177, 1963.

[31] Drayson, M.: "An Amplitude Distribution Meter", *Electr. Eng.,* vol. 31, no. 10, pp. 578—584, 1959.

[32] Fränz, K.: "Die Empfindlichkeit bei Schreibempfang und Instrumentenbeobachtung", *Hochfrequ. u. El. ak.,* vol. 58, no. 10, pp. 95—99, 1941.

[33] Fritzsche, G.: "Netzwerke mit vorgeschriebener Wurzelcharakteristik in einem breiten Frequenzband", *Hochfrequ. u. El. ak.,* vol. 72, no. 3, pp. 87—93, 1963.

[34] Güttler, P.: "Rauschmessungen bei tiefsten Frequenzen", *Nachrichtentechnik,* vol. 19, no. 8, pp. 287—291, 1969.

[35] —: "Ein Messplatz zur Messung tiefstfrequenten Rauschen von Transistoren", *Nachrichtentechnik,* vol. 20, no. 2, pp. 69—75, 1970.

[36] Haus, H. A. *et al.*: "IRE Standards on Methods of Measuring Noise in Linear Two-Ports", *Proc. IRE,* vol. 48, no. 1, pp. 60—68, 1960.

[37] Huelsman, L. P.: *Active Filters,* McGraw-Hill, New York, 1970.

[38] Itoh, H. and K. L. Knudsen: "Direct Measurement of Transistor Noise Voltage, Noise Current and Noise Figure", *Hewlett Packard J.,* vol. 21, no. 2, pp. 2—7, 1969.

[39] Kitai, R. and D. Braithwaite: "Digital Statistical Analyser for Low Frequencies", *Electr. Lett.*, vol. 4, no. 10, pp. 196—197, 1968.

[40] Knott, K. F.: "Measurement of Battery Noise and Resistor-Current Noise at Subaudio Frequencies", *Electr. Lett.*, vol. 1, no. 5, p. 132, 1965.

[41] Kovács, F. and A. Pócza: "Automatische Messvorrichtung zur Bestimmung des HF Rauschfaktors von Transistoren", *Int. Elektr. Rdsch.*, vol. 25, no. 4, pp. 89—92, 1971.

[42] Lien, H.: "Probability Density Measurement with an Electrode Mounted in the Face of a Cathode Ray Tube", *Rev. Sci. Instr.*, vol. 30, no. 12, pp. 1100—1102, 1959.

[43] López de la Fuente, J.: "Measurement of Very Low Frequency Noise", Doctor's dissertation, University of Eindhoven, 1970.

[44] Mamola, G. and M. Sanino: "Considerations on the Measurement of Linear Two-ports, Noise Parameters", *Alta Frequenza*, vol. 42, no. 10, pp. 535—544, 1973.

[45] Mansour, I. R. M. *et al.*: "Digital Analysis of Current Noise at Very Low Frequencies", *Radio Electr. Eng.*, vol. 35, no. 4, pp. 201—211, 1968.

[46] Martin, A. J.: Private communication, May 20, 1970.

[47] Merlo, D. *et al.*: "Effect of Some Component Tolerances and Measuring Errors on Noise Measurement", *Electr. Lett.*, vol. 1, no. 3, p. 250, 1965.

[48] Neuvo, Y. and W. H. Ku: "Analysis and Digital Realization of a Pseudorandom Gaussian and Impulsive Noise Source", *IEEE Trans.*, vol. COM-23, no. 9, pp. 849—858, 1975.

[49] Newland, D. E.: *Random Vibrations and Spectral Analysis*, Longman, London, 1975.

[50] Nikiforuk, P. N. and G. Squires: "A Technique for Probability Density Function Measurement", *Electr. Eng.*, vol. 37, no. 5, pp. 316—317, 1965.

[51] Ojala, L. and E. T. Rautanen: "On Incremental Algorithms for Averaging and Correlation Computation", *IEEE Trans.*, vol. IM-23, no. 1, pp. 90—94, 1974.

[52] Pfeiler, M.: "Über den Einfluss der Verbindungsleitung zwischen Rauschgenerator und Vierpol bei Rauschzahlmessungen", *NTZ*, vol. 18, no. 9, pp. 531—537, 1965.

[53] Rex, R. L. and G. T. Roberts: "Correlation, Signal Averaging and Probability Analysis", *Hewlett Packard J.*, vol. 21, no. 3, pp. 2—8, 1969.

[54] Rheinfelder, W. A.: *Design of Low-Noise Transistor Input Circuits*, Hayden, New York, 1964.

[55] Rice, S. O.: "Mathematical Analysis of Random Noise", *BSTJ*, vol. 23, no. 3, pp. 228—233, 1944; vol. 24, no. 1, pp. 46—157, 1945.

[56] Rubin, L. G. and Y. Golahny: "Comment on a Simple Analogue Device for Constructing the Time Average of a Fluctuating Input", *J. Phys.*, vol. E8, no. 12, p. 977, 1975.

[57] Samways, P. R.: "Optimum Noise Filter for DC Measurements", *J. Phys.*, vol. E1, no. 2, pp. 142—144, 1968.

[58] Smith, D. T.: "A Linear Detector and Integrator for Low-Frequency Noise Measurements", *J. Phys.*, vol. E3, no. 6, pp. 472—474, 1970.

[59] Soeberg, J.: "Real Time Analysis", *Brüel-Kjaer Tech. Rev.*, no. 4, pp. 3—11, 1969.

[60] Solartron Data Sheet: "Time Domain Analyser JM 1860".

[61] Strasilla, U. J. and M. J. O. Strutt: "Measurement of White and $1/f$ Noise Within Burst Noise", *Proc. IEEE*, vol. 62, no. 12, pp. 1711—1713, 1974.

[62] Sutcliffe, H.: "Noise Spectrum Measurement at Subaudio Frequencies", *Proc. IEE*, vol. 112, no. 2, pp. 301—309, 1965.

[63] —: "Mean Detector for Slow Fluctuations", *Electr. Lett.*, vol. 4, no. 6, p. 97, 1968.

[64] —: "Relative Merits of Quadratic and Linear Detectors in the Direct Measurement of Noise Spectra", *Radio Electr. Eng.*, vol. 42, no. 2, pp. 65—68, 1972.

[65] — and G. H. Tomlinson: "A Low-Frequency Gaussian White-Noise Generator",. *Int. J. Control*, vol. 8, no. 5, pp. 457—471, 1968.

[66] Székely, V.: "Technical Applications of Deconvolution"; (A dekonvolúció technikai alkalmazásai), *Híradástechnika*, vol. 22, no. 6, pp. 169—177, 1971, (in Hungarian).

[67] Temes, G. C. and S. K. Mitra: *Modern Filter Theory and Design*, John Wiley, New York, 1973.

[68] Tiuri, M. E. and S. J. Halme: "Digital Measurement of Narrow-Band Noise Power", *Proc. IEEE*, vol. 55, no. 9, pp. 1577—1583, 1967.

[69] Upton, R.: "An Objective Comparison of Analog and Digital Methods of Real Time Frequency Analysis", *Brüel–Kjaer Tech. Rev.*, no. 1, pp. 18—26, 1977.

[70] Wehrmann, W.: *Einführung in die stochastisch-ergodische Impulstechnik*, R. Oldenbourg, Vienna, 1973.

[71] Yang, E. S.: "A Probability Density Analyzer", *IEEE Trans.*, vol. IM-18, no. 1, p. 15, 1969.

[72] The Fast Fourier Transform, *IEEE Trans.*, vol. AU-15, no. 2, p. 43, 1967.

SYMBOLS

In this list, only frequently used notations are given to help in quick location. Notations used in short derivations, or infrequently used notations, are better found in the text. Equations where notations first appear are given.

a	stochastic amplitude (4-101)
a	real part of correlation factor (5-46)
a	coefficient of linear characteristic: $y=a\lvert x\rvert$ (7-10)
a_{ij}	elements of chain matrix (5-83)
a_0	twice the DC term in Fourier series (3-3)
a_n	coefficients of cosine terms in Fourier series (3-3)
A	cross section (4-44)
A	large signal current gain factor of transistor operating in common base configuration (6-42)
A_p	power gain of two-port (5-39)
b	$=\mu_n/\mu_p$, the ratio of mobilities (4-69)
b	imaginary part of correlation factor (5-46)
b	coefficient of square-law characteristic: $y=bx^2$ (7-7)
b_n	coefficients of sine terms in Fourier series (3-3)
B	large signal current gain factor of a transistor operating in common emitter configuration (6-63)
B_c, B_f, B_w, B_n	noise bandwidth designations (5-91) to (5-97)
c	complex correlation factor (5-32)
c, C	constant additive term or multiplying factor (2-18)
c_n	$=(a_n-jb_n)/2$, complex coefficients of the Fourier series (3-12)
C_{gc}	gate–channel capacity of field-effect transistor (6-141)
$C_{\xi\eta}$	covariance of random variables ξ and η (3-43)
d	distance between electrodes (4-2)

d_i	width of insulating layer (6-115)
D_n, D_p	diffusion constants (6-2)
$D^2(\ )$	variance of the random variable in parentheses (1-6)
E	field strength (4-3)
f	frequency (1-11)
f_a	cut-off frequency of low-pass filter (5-92)
f_c	crossover frequency of white noise and flicker noise (5-94)
f_0	band-center frequency (5-91)
f_α	cut-off frequency of transistor current gain factor α (6-52)
$f(n), f(x)$	density function (1-1)
F	force (4-3)
F	noise figure (5-39)
$F(x)$	distribution function (2-2)
$F(\omega)$	Fourier transform of time function $x(t)$ (3-21)
$F_T(\omega)$	Fourier transform of truncated time function $x_T(t)$ (3-24)
$F_\Gamma(\omega)$	Fourier transform of autocorrelation function $\Gamma(\tau)$ (3-56)
g_m	mutual conductance (6-59)
g_{ms}	mutual conductance of a field-effect transistor in the pinch-off region (6-123)
G	conductance (4-51)
G_e	real part of emitter–base admittance of transistor operating in common base configuration (6-38)
G_{e0}	low-frequency value of G_e (6-38)
G_n	equivalent noise conductance (5-38)
G_0	channel conductance of junction field-effect transistor at zero electrode voltages (6-116)
h	$=6.62\times10^{-34}$ Js, the Planck constant (4-54)
h	error (8-9)
$h(\ )$	weight function of linear, time-invariant network (3-106)
$H(\ )$	transfer function of linear, time-invariant network (3-111)
i	instantaneous value of alternating current without DC component, or instantaneous value of deviation from mean value (1-9)
i	instantaneous value of current comprising DC component (1-7)
i_b	base-noise current of transistor (6-57)
i_d	drain-noise current of field-effect transistor (6-129)
i_g	gate-noise current (6-134)
i_0	source current of equivalent noise generator (5-2)
i_1, i_2	input and output currents of two-port (5-24)
i_1	emitter-noise current (6-33)
i_2	collector-noise current (6-34)
i'_2	noise current originating in the collector circuit (6-39)
I_B	base current (6-43)

I_{CB0}	collector–base leakage current (6-42)
I_C	collector current (6-34)
I_D	drain current of field-effect transistor (6-116)
I_{DS}	saturation drain current of field effect transistor (6-121)
I_E	emitter current (6-33)
I_{gen}	generation current (6-30)
I_n	electron current (4-56)
I_p	hole current (4-56)
I_R	recombination current (6-27)
I_0	reverse current of p–n junction (6-1)
j	$\sqrt{-1}$
J	current density (4-42)
k	$=1.38 \times 10^{-23}$ J/K, the Boltzmann constant (4-23)
$K_d(\eta)$, $K_g(\eta)$, $K_{dg}(\eta)$	working point-dependent factors, characterizing the channel noise and gate noise of a MIS transistor, and their correlation (6-131), (6-141), and (6-145)
$K_d(x, y)$, $K_g(x, y)$, $K_{dg}(x, y)$	working point-dependent factors, characterizing the channel noise and gate noise of a p–n junction field-effect transistor, and their correlation (6-132) and Fig. 6-28
l	length (4-44)
L	chain matrix (5-30)
m	factor characterizing recombination current ratio (6-28)
m	(as superscript) depending on dopant profile (6-114)
m^*	effective mass of electron or hole (4-38)
$M(\ )$	expected value of the bracketed random variable (1-5)
M	network matrix (5-23)
$M(x)$	multiplying factor (4-56)
n, N	number of pieces (1-1)
n	(as subscript) serial number of term in a series (2-19)
n_0	number of charge carriers per unit volume (4-42)
N	number of *all* electrons within a given volume (4-68)
p	power density (1-17)
P	power (1-16)
P	number of *all* holes within a given volume (4-68)
p_i	occurence probability of the event $\xi = x_i$ (2-10)
$\mathscr{P}[\ ]$	probability of the bracketed event (1-13)
P_{ni}	two-port input noise power (5-39)
P_{n0}	two-port output noise power (5-39)
P'_{n0}	part of two-port output noise power originating within the two-port (5-39)
q	$=1.6 \times 10^{-19}$ As, the charge of the electron (1-7)
q_g	noise component of gate charge (6-141)

V_T	threshold voltage of MOS-transistor (6-115)
w	width of depletion layer (4-56)
W	energy (4-3)
x	set of permitted values of random variable ξ (2-4)
$x(t)$	deterministic time function (3-3)
$x_T(t)$	truncated time function which is zero outside the interval $\pm T$ (3-23)
x, y	factors characterizing working point adjustment of p–n junction field-effect transistor (6-132)
y	set of permitted values of random variable η (2-48)
$y(t)$	response to the driving function $x(t)$ (3-106)
$\mathbf{Y}$	admittance matrix (5.24)
Y_c	$=G_c+jB_c$, correlation admittance (5-33)
Y_g	$=G_g+jB_g$, admittance of signal source (5-56)
Y_{21}	complex g_m of transistor (6-36)
z	random variable with zero expected value normalized to σ (2-54)
$\mathbf{Z}$	impedance matrix (5-71)
Z_g	$=R_g+jX_g$, internal impedance of signal source (5-44)
Z_p	parallel impedance (5-86)
Z_s	series impedance (5.85) (6-24)
$1(x)$	unit step function (7-12)
I, II	(as superscripts) designation for noise generators excluded from two-port (5-23)
$*$	(as superscripts) conjugated complex number or function (2-41)
α	$= 2\times10^{-3}$, Hooge constant (4-93)
α	complex current gain factor of transistor operating in common base configuration (6-35)
α_0	low-frequency, small signal current gain factor of transistor operating in common base configuration (6-47)
β	small signal current gain factor of transistor operating in common emitter configuration (6-79)
β_0	low-frequency, small signal current gain factor of transistor operating in common emitter configuration (6-79)
γ	conductivity (4-43)
γ	coefficient in equation expressing the current–voltage characteristic of a MIS field-effect transistor (6-118)
$\gamma(t)$	normalized autocorrelation function (7-17)
$\Gamma_\xi(\tau)$	autocorrelation function of the stochastic process $\xi(t)$ (3-2)
$\Gamma_{\xi\eta}$	correlation of random variables ξ and η (3-45)
$\Gamma_{u_1 u_2}$	crosscorrelation function of stochastic processes $u_1(t)$ and $u_2(t)$ (5-9)
$\delta(\omega)$	Dirac-delta function (3-78)

$\delta(I), \delta(U), \delta(Q)$	perturbation of current, voltage, charge (6-126)
ε_i	relative dielectric constant of insulator (6-110)
ε_0	$= 8.85 \times 10^{-12}$ F/m, permittivity of the vacuum (6-110)
$\zeta, \zeta(t)$	transformed random variable (7-32)
$\eta, \eta(t)$	sum or transformed random variable (2-44)
η	factor characterizing the working point adjustment of a MIS transistor (6-130)
ϑ	auxiliary variable of time dimension (3-50)
$\vartheta(t)$	phase angle of stochastic variable (2-131)
Θ	auxiliary variable of time dimension (3-108)
Θ	average time interval between two impacts (4-34)
λ	expected value of Poisson distribution (1-2)
λ	parameter of exponential distribution (2-144)
μ_n	n-th order moment of the distribution (2-36)
μ_n, μ_p	mobilities of charge carriers (4-68)
ν	number of state changes per time unit (3-100)
$\nu(t)$	stochastic time function (7-18)
ν_l, ν_s	relative fluctuation of signal after linear or square-law detection and RC low-pass filtering (8-17)
$\xi, \xi(t)$	random variable (Sec. 2-1)
$\varrho, \varrho(t)$	stochastic time function of envelope (2-131)
$\varrho_x(\tau)$	correlation parameter of the gaussian process (3-74)
ϱ	$= r/r_k$, normalized radius (4-14)
σ	standard deviation of normal distribution (2-52)
τ	variable of autocorrelation function (time delay) (3-2)
φ	phase angle (1-14)
$\varphi(v)$	characteristic function (2-37)
χ^2	squared value of random variable with normal distribution, or sum of squared values of several random variables with normal distribution (2-117)
Ψ_l, Ψ_s	relative standard deviation of signal originating from linear or square-law detection and ideal integration (8-5)
ω	angular frequency (2-131)
$\Delta\omega$	bandwidth (2-131)
Ω	auxiliary variable having angular frequency dimension (7-26)
Ω	event space of probability theory (2-2)

SOME DEFINITE INTEGRALS*

$$\int_0^\infty e^{-ax^2} \cos bx \, dx = \frac{1}{2} \sqrt{\frac{\pi}{a}} \, e^{-b^2/4a} \tag{B-1}$$

$$\int_0^\infty \left(\frac{\sin x}{x}\right)^n \cos mx \, dx = \begin{cases} \dfrac{n\pi}{2^n} \displaystyle\sum_{k=0}^{(m+n)/2} \dfrac{(-1)^k (n+m-2k)^{n-1}}{k!(n-k)!}, & \text{if} \quad 0 < m < n \\[2ex] 0, & \text{if} \quad m \geqq n \quad \text{and} \quad n \geqq 2 \end{cases} \tag{B-2}$$

$$\int_0^\infty \frac{\cos ax}{b^2 + x^2} \, dx = \frac{\pi}{2b} \, e^{-|ab|} \tag{B-3}$$

$$\int_0^\infty e^{-px} \cos (qx + \lambda) \, dx = \frac{1}{p^2 + q^2} (p \cos \lambda - q \sin \lambda) \tag{B-4}$$

* Gradshtein, I. S. and I. M. Ryzhik, *Table of Integrals, Series, and Products*, Academic Press, New York, 1965.

APPROXIMATION OF $n!$*

This approximation is more accurate than the Stirling formula, and is easily programmable.

$$n! \cong (2\pi n)^{1/2} \, e^{(n[\ln (n) - 1] + n/12)} \qquad \text{(C-1)}$$

* *Electronics*, vol. 50, no. 20, p. 114, 1977.

INDEX